THE BAREBONES GUIDE SAT MATH LEVEL 2

Tutor Associates Mobile App Inc., Wilmington, DE 19801

ISBN-13: 978-0692391914

ISBN-10: 0692391916

Cover design and interior artwork by Gabrielle Harrison

CONTENTS

PREFACE

The Barebones Guide is in part the product of our frustration with commercially available SAT Subject Test preparation materials—poorly written, irritatingly cheerful books offering unrealistic and unhelpful advice. This book doesn't pretend that taking standardized tests is a wacky good time, and it doesn't expect you to learn a year's worth of trigonometry and precalculus in a month (note, however, that our tests don't skip these topics—they are, if anything, somewhat harder than the official tests). Instead, we have tried to provide a simple approach based on what we tell our own students: get familiar with the format of the test, make sure you know a few key concepts, hack questions you can't solve directly, and skip anything you aren't sure about.

The order in which we have presented the content of this book is not intended as a prescription for studying. Perhaps you will prefer to take a test before looking at any advice, or vice versa. Perhaps you are only interested in the tests or only in strategy advice. These are all legitimate approaches to studying. In our experience, students improve the most with some combination of testing and reviewing strategy and content. Taking a test and then analyzing how you answered each question (and perhaps what you would do differently a second time around) is often helpful. With that said, we leave it to you.

PART ONE: ADVICE

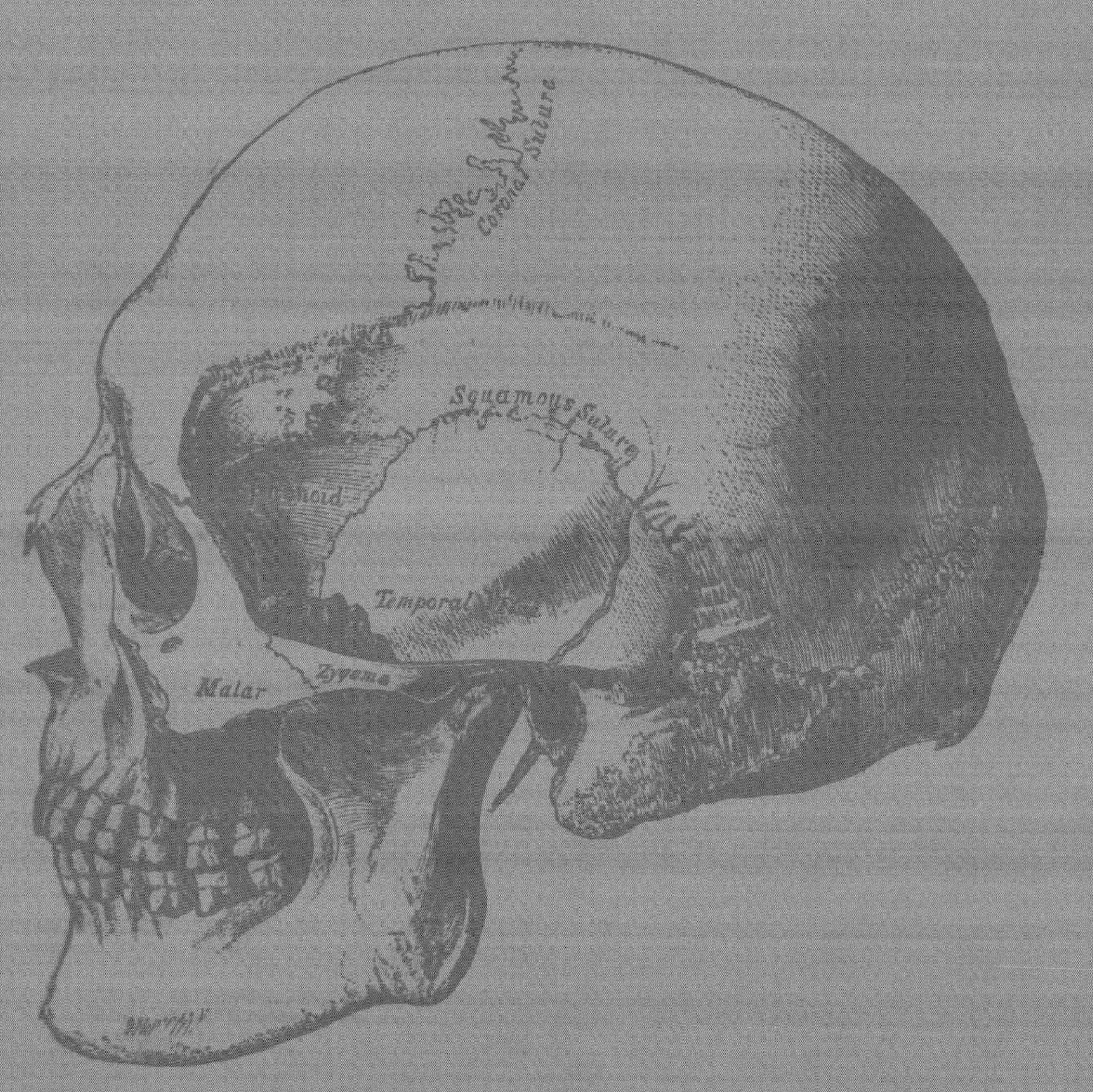

ABOUT THE SAT MATH LEVEL 2

The SAT Subject Test in Mathematics Level 2[1] is written to cover topics you have already encountered in your high school math classes. However, it differs from tests you have taken for school in a few notable ways:

You do not need to answer every question.

- While the curve varies from year to year, you can generally skip six to seven questions and still score an 800 (if you answer the other questions correctly, that is). For a 750, you can skip around twelve questions; for a 700, seventeen; for a 650, twenty-two; and for a 600, twenty-eight.[2] Use this leeway to your advantage: only answer questions you are confident about.

Wrong answers are counted against you.

- There is a small guessing penalty incurred for each incorrect answer you enter. Don't stress about this—instead, take it as an incentive to skip questions you aren't sure about.[3]

There is no partial credit.[4]

- The flip side of this is that no one is checking your work, so any way you can narrow down your answers is legit.[5]

Since the SAT Math Level 2[6] differs from tests you have taken for school, you will probably want to prepare for it differently than you would for a school test. These are our recommendations:

Practice skipping

- Easy questions count for just as much as hard questions, so skip the hard questions the first time around—if you're spending more than a couple minutes on a question, drop it and move on.[7] You can come back to any question you've skipped.

---

1 The official, unnecessarily long name for windbags.
2 Pretty nice, right?
3 Skipping = less work and a better score. What's not to love?
4 Okay, so maybe you've taken a test without partial credit before. We just wanted to be thorough.
5 Necromancy excluded.
6 Isn't that so much better? Short and sweet.
7 Let it gooooo, let it gooooo....

Get cozy with your calculator

- Your calculator can be your best friend or your worst enemy.[8] We encourage you to get comfortable with the following:
  - graphing functions
  - finding mins, maxes, and zeros of functions
  - calculating and manipulating logs and exponents
  - summing sequences
  - calculating basic trig functions

Use guerrilla mathematics

- Are the answer choices numbers? See if you can plug them into an equation in the question.
- Are they variables? Come up with a few sample numbers and try them out to eliminate wrong answers.
- Is there something you can draw? Draw it.[9]
- Use. Your. Calculator.[10]

---

8 I made friends with my calculator, and now my worst enemy is bacne. Go figure.
9 Always draw. It has saved me many a time.
10 It bears repetition.

ADVICE BY TOPIC

-OR-

HOW TO APPEAR TO KNOW THINGS

TRIGONOMETRY

Use your calculator

- Know when to be in degree mode and when to be in radian mode.[1]
- Calculate values when you can. If this seems obvious, that's because it is, but it's still good to practice.
- Convert everything to sines and cosines (from sec, csc, etc.) before plugging in.

Hack questions with guerrilla math

- If answer choices are numbers, plug them into your calculator and see which one works.
- If your answer choices are variables, use a hypothetical angle like 30 degrees and plug in to eliminate wrong answers.
  - It's very helpful to know that sin30= ½, cos60 = ½, and tan45 = 1.[2]

Don't waste your time

- Almost every trig identity is a waste of time.
- If you're not sure how to do a question directly and your calculator and guerilla math aren't helping, skip it.

FUNCTIONS

Use your calculator

- The TABLE function on the calculator is your secret weapon,[3] especially on domain and range questions.
- ZERO and MAX/MIN are also good.[4]

Hack questions with guerrilla math

- Making up a value for x and plugging it in can be helpful, especially for inverse function problems.

---

1 If you do the whole test in degree mode, you're gonna have a bad time.
2 So spend two minutes memorizing that.
3 So secret, in fact, that you may not have known about it. Now you do. You're welcome.
4 But they lack a certain je ne sais quoi, if you know what I mean.

Don't waste your time

- Factoring and synthetic division are not worth your time on this test.[5] Use your calculator instead.
- If you're not sure how to do the question directly and your calculator and guerrilla math aren't helping, skip it.

LOGS AND EXPONENTIALS

Use your calculator

- Remember the log button is base 10 and the ln button is base e.[6] Get comfortable calculating logs with various bases with your calculator.

Hack questions with guerrilla math

- If a question looks complicated, try converting between log and exponential form.
- Have an easy example handy, like $\log_3 9 = 2$, to help you plug in your own example.
- This may not exactly qualify as guerrilla math, but it is very helpful to know your exponent rules.[7]

Don't waste your time

- Don't spend time memorizing the change of base rule for logs if your calculator can calculate any logarithm directly.[8] Otherwise it's good to know.
- If you're not sure how to do the question directly and your calculator and guerrilla math aren't helping, skip it.[9]

---

5 They're also super boring, amirite?
6 If you're curious about e, there's a great book by Eli Maor on the subject.
7 Quick, what is the difference between $(x^3)(x^2)$ and $(x^3)^2$? That type of thing.
8 The newer TI-84 model can do this. Some other calculators can't.
9 It cannot be said too frequently or forcefully: skip, skip, skip!

LOWER-HANGING FRUIT

ODDS AND ENDS TO MEMORIZE

A LITTLE BIT O' TRIG

Ideally, you are already familiar with SOHCAHTOA, or it at least rings a bell. If not, go look it up.

Take a moment to absorb (i.e. make flashcards for) the information in these charts:

QUADRANT II: SINE Only sine is positive	**QUADRANT I: ALL** Sine, cosine, and tangent are all positive
QUADRANT III: TANGENT Only tangent is positive	**QUADRANT IV: COSINE** Only cosine is positive

$$\csc x = \frac{1}{\sin x}$$

$$\sec x = \frac{1}{\cos x}$$

$$\cot x = \frac{1}{\tan x}$$

$$\arcsin x = \sin^{-1} x$$

$$\arccos x = \cos^{-1} x$$

$$\arctan x = \tan^{-1} x$$

A FEW CONICS

$(x - h)^2 + (y - k)^2 = r^2$ is a circle with center (h, k) and radius r.

$\frac{(x-h)^2}{a^2} + \frac{(y-k)^2}{b^2} = 1$ is an ellipse with center (h, k).

$\frac{(x-h)^2}{a^2} - \frac{(y-k)^2}{b^2} = 1$ is a hyperbola with center (h, k).

A WORD OR TWO ABOUT FUNCTIONS

If $f(g(x)) = x$, then f and g are inverses of each other.

To find the inverse of a function, switch x and y, and solve for (the new) y.

The domain of a function is what x-values it takes as input (how far right and left its graph goes, minus any holes in between).

The range of a function is what y-values it produces as output (how high and low its graph goes, minus any holes in between).

TAKING THE TEST

A day or two before your test, you should switch gears: stop studying and focus on getting getting ready for the test mentally and physically. Your priority should be to feel rested, calm, and organized on the day of the test.[1] Here's the checklist we use:

Take care of yourself

- Don't study the night before the test. Instead, do something relaxing. Read a book, watch a TV show, bake a pie—whatever floats your boat.[2] If you feel you absolutely must study something, choose one or two limited topics or strategies to review for a half hour.[3]
- Get enough sleep—need we say more?
- Eat a healthy breakfast. Oatmeal with fruit is a good bet; sugary cereals, poptarts, etc. are not.[4]

Get prepared ahead of time

- Last-minute panic is not great for your nerves,[5] so pack everything you need for your test the night before:
    - Your admission ticket
    - Several number 2 pencils (not mechanical, which are not permitted)[6]
    - Your calculator and perhaps some back-up batteries
    - A bottle of water
    - A snack—bananas, almonds, and energy bars are all good. Keep in mind that even if you're only taking one test, you may have to wait a few hours at your test site before your test begins.

Stick to your guns during the test

- Remember that you should expect to skip questions on this test—if you don't know how to answer a question, just don't answer it.[7]
- If you've taken practice tests using a specific game plan,[8] use that plan when you take the real test.

---

1 Unless, of course, you perform best when you're exhausted, distraught, and coming apart at the seams. You do you.

2 I read *Goodnight Moon* before tests, but that's just me.

3 And no more.

4 Before a test, or really ever, but I digress.

5 Or at least for mine.

6 Not that I've ever brought the wrong kind of pencil to a test.

7 Just skip it. No sweat.

8 We encourage this.

PART TWO: TESTS

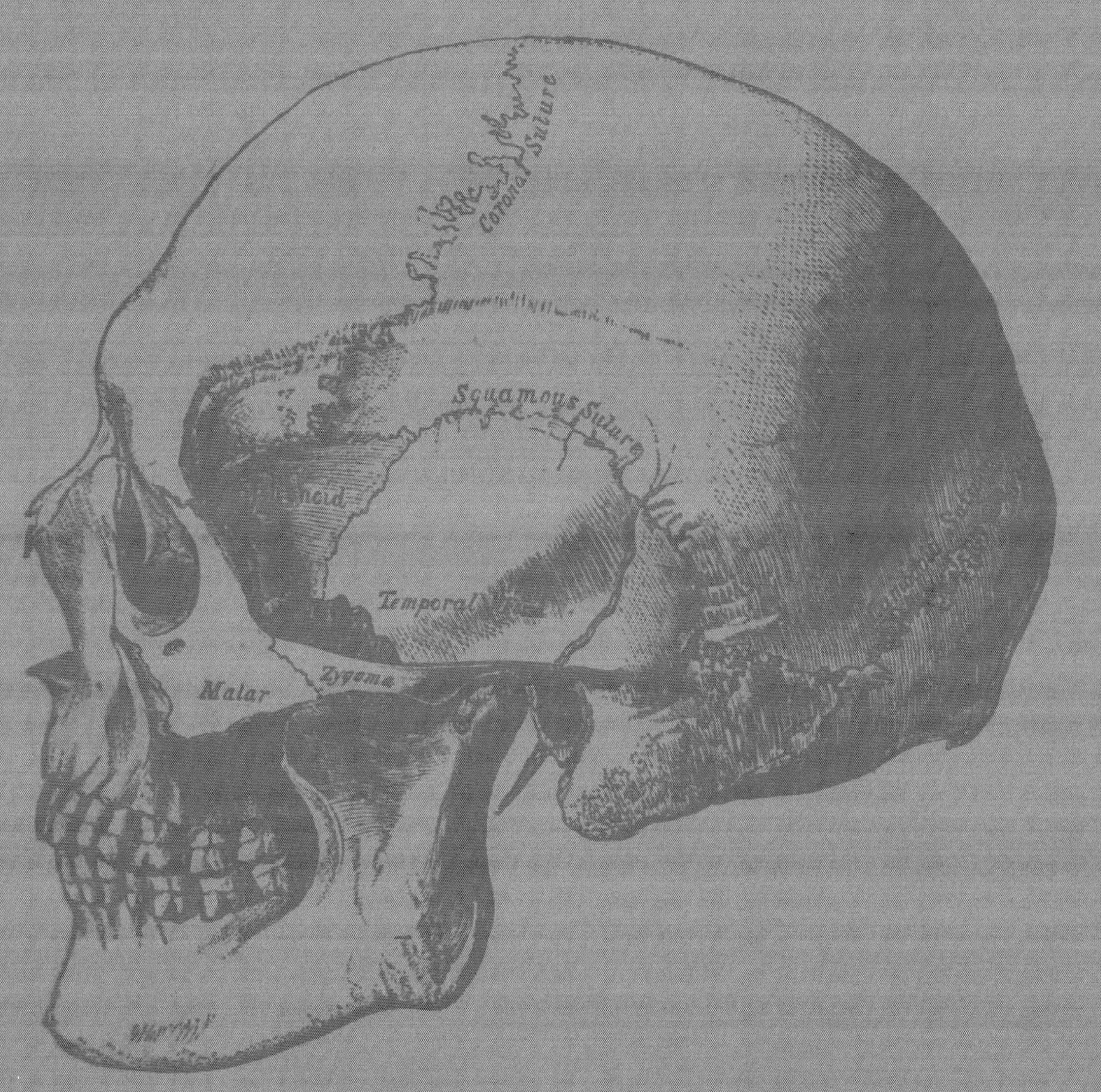

TEST ONE
MANDATORY FUN

For most accurate results, take this test in a setting similar to the one you'll take the real test in. We've organized the essentials into this simple mnemonic:

Choose a quiet spot free from distractions.

Remember to use a scientific or graphing calculator.

Outside references are a no-no.

Use an answer sheet (in the back of this book or downloaded from barebonesguide.com/answersheet) to record your answers.

Place a timer within sight and set it to one hour.

After you have finished, consult our scoring instructions at the end of the test to calculate your raw and scaled scores.

MATHEMATICS LEVEL 2 TEST

REFERENCE INFORMATION

YOU MAY USE THE FOLLOWING INFORMATION AS YOU TAKE THIS TEST.

A right circular cone with radius r and height h has volume $V = \frac{1}{3}\pi r^2 h$

A right circular cone with circumference of the base c and slant height ℓ has lateral area $S = \frac{1}{2}c\ell$

A sphere with radius r has volume $V = \frac{4}{3}\pi r^3$

A sphere with radius r has surface area $S = 4\pi r^2$

A pyramid with base area B and height h has volume $V = \frac{1}{3}Bh$

GO ON TO THE NEXT PAGE

MATHEMATICS LEVEL 2 TEST

Choose the best choice given for each of the following questions. The exact value may not be one of the choices; in this case select the best approximation of the true value.

Note the following items carefully.

(1) You will need a scientific or graphing calculator to answer some of the questions in this test. Some of these questions will require you to determine whether your calculator should be in degree or radian mode. For other questions, however, you will not need a calculator at all.

(2) All figures are drawn to scale and lie in the plane unless otherwise specified.

(3) For the purposes of this test, the domain of any function f is the set of real numbers x for which $f(x)$ is a real number and the range of f is the set of real numbers $f(x)$, where x is in the domain of f. Any deviation from these assumptions will be noted in each problem to which it applies.

USE THIS SPACE FOR SCRATCHWORK.

1. If $2 - \frac{4}{x} = 6 - \frac{12}{x}$, then $2 - \frac{4}{x} =$?

(A) -2
(B) -1
(C) 0
(D) 1
(E) 2

2. $x(\frac{3}{y} + \frac{5}{z}) =$

(A) $\frac{15x}{yz}$
(B) $\frac{15x}{y+z}$
(C) $\frac{8x}{y+z}$
(D) $\frac{5xy + 3xz}{yz}$
(E) $\frac{15}{xy + xz}$

GO ON TO THE NEXT PAGE

USE THIS SPACE FOR SCRATCHWORK.

3. What is the slope of the line containing points $(7, 2)$ and $(3, -6)$?

   (A) -13
   (B) -2
   (C) 1
   (D) 2
   (E) 13

4. What is the degree measure of one of the smaller angles of a triangle with side lengths 15, 8 and 8?

   (A) $20°$
   (B) $41°$
   (C) $78°$
   (D) $139°$
   (E) $145°$

5. The formula $A = Pe^{0.06t}$ gives the amount A that a savings account will be worth after an initial investment P is compounded continuously at an annual rate of 6 percent for t years. Under these conditions, what is the approximate initial amount that should be invested to have a final balance of \$20,000 after 10 years?

   (A) $\$9,000$
   (B) $\$10,000$
   (C) $\$11,000$
   (D) $\$12,000$
   (E) $\$13,000$

GO ON TO THE NEXT PAGE

USE THIS SPACE FOR SCRATCHWORK.

6. The relationship between one BTU on the British thermal unit scale and joules is 1 BTU = 1,055 joules, and the relationship between one horsepower and joules per second is 1 horsepower = 745 joules per second. Which of the following represents the relationship between horsepower and British thermal units per second?

   (A) 1 horsepower = 0.706 BTU per second
   (B) 1 horsepower = 1.000 BTU per second
   (C) 1 horsepower = 1.416 BTU per second
   (D) 1 horsepower = 745 BTU per second
   (E) 1 horsepower = 1,055 BTU per second

7. If $f(x) = \sqrt{2x^2 + 0.5x}$ and $g(x) = \dfrac{x-1}{x+2}$, then $g(f(2))=$

   (A) 0.25
   (B) 0.4
   (C) 0.5
   (D) 1
   (E) 1.5

8. The graph of which of the following equations has an undefined slope?

   (A) $x = 0$
   (B) $x = y$
   (C) $y = -x$
   (D) $y = 0$
   (E) $y = 1$

GO ON TO THE NEXT PAGE

USE THIS SPACE FOR SCRATCHWORK.

9. A balloon on a string, tied at ground level, is blowing in the wind. If the balloon is 12 feet from the ground and the string makes an angle of 39° with the ground, how long is the string?

(A) 8 feet
(B) 9 feet
(C) 15 feet
(D) 19 feet
(E) 23 feet

10. If $4x + 12 = \frac{k}{5}(x + 3)$ for all x, then $k =$

(A) 0.2
(B) 1
(C) 4
(D) 5
(E) 20

GO ON TO THE NEXT PAGE

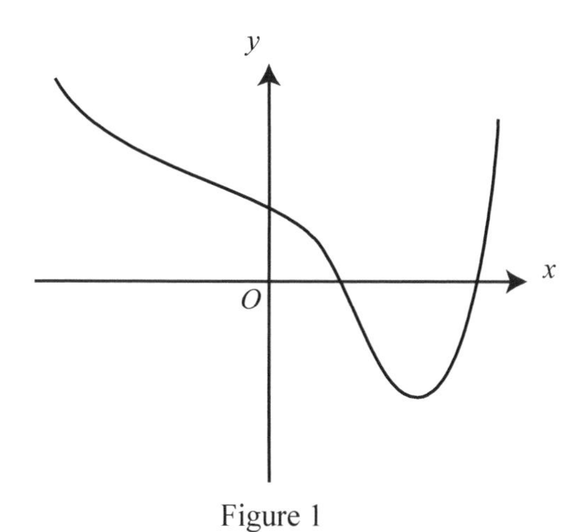

Figure 1

11. The graph of $y = f(x)$ is shown in Figure 1. Which of the following could be the graph of $y = |f(x)|$?

(A)
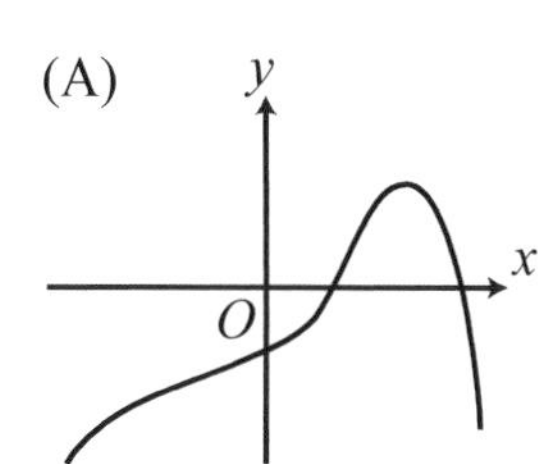

(B)
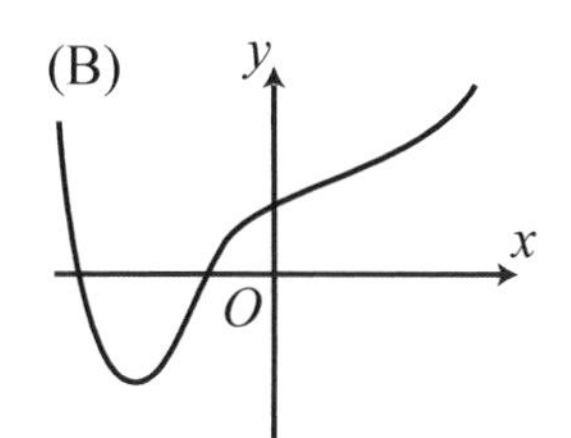

(C)
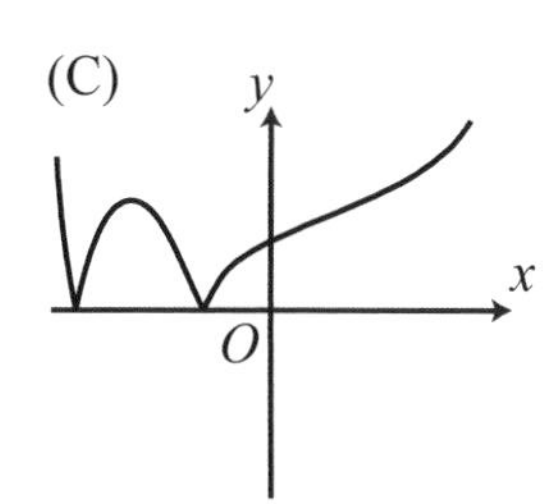

(D)
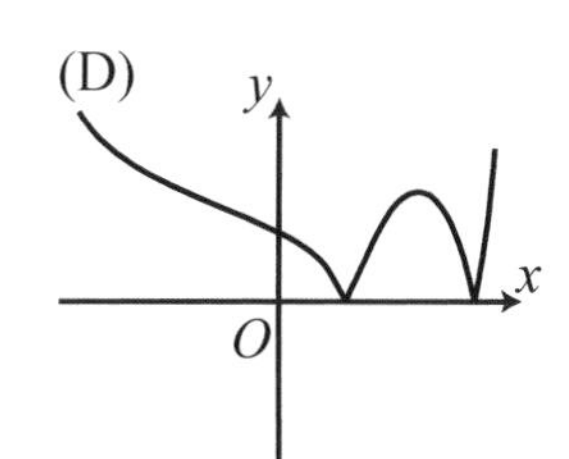

(E)
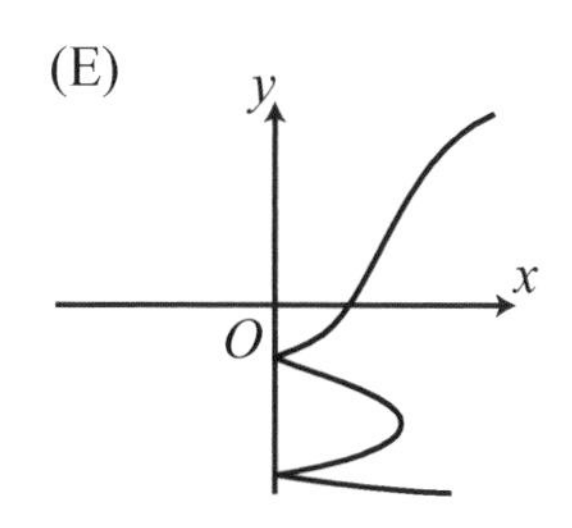

GO ON TO THE NEXT PAGE

USE THIS SPACE FOR SCRATCHWORK.

12. If $f^{-1}(x) = e^{(x^2)}$, then $f(x) =$

(A) $2 \ln x$

(B) $\ln 2x$

(C) $\ln \sqrt{x}$

(D) $\ln \frac{x}{2}$

(E) $\sqrt{\ln x}$

13. On the interval $0 \leq x < 2\pi$, where does the graph of $y = \cos x$ attain a minimum?

(A) $(0, 0)$

(B) $(\pi, -1)$

(C) $(\frac{\pi}{2}, -1)$

(D) $(0, -1)$

(E) $(\frac{3\pi}{2}, -1)$

14. If $f(x) = 6x - 4$ and $f(g(2)) = 26$, which of the following could be $g(x)$?

(A) $3x + 1$
(B) $2x - 3$
(C) $2x + 1$
(D) $x + 2$
(E) $3x + 2$

GO ON TO THE NEXT PAGE

USE THIS SPACE FOR SCRATCHWORK.

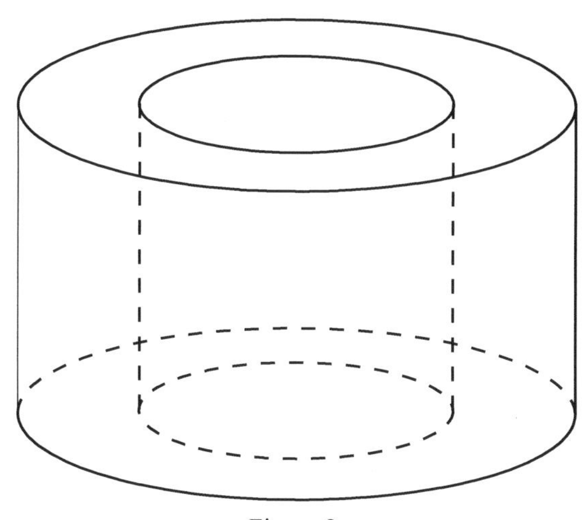

Figure 2

15. The figure above shows a hollowed out cylinder with an inner and outer radius of r and $2r$, respectively. Which expression gives the volume of the solid?

(A) $\frac{4\pi r^2 h}{3}$

(B) $3\pi rh$

(C) $3\pi r^2 h$

(D) $\frac{3\pi r^2 h}{4}$

(E) $5\pi rh$

GO ON TO THE NEXT PAGE

USE THIS SPACE FOR SCRATCHWORK.

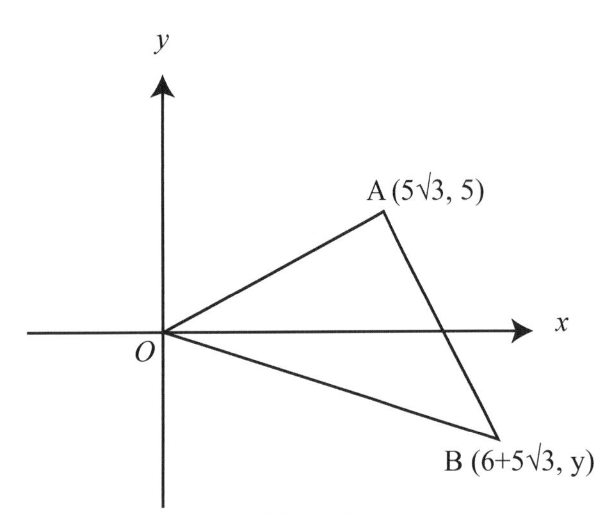

Figure 3

16. In Figure 3, O is the origin. If $\overline{AB}$ has a slope of $-\frac{4}{3}$, what is the value of y?

    (A) $-5\sqrt{3}$
    (B) $-3+\sqrt{3}$
    (C) -4
    (D) -3
    (E) It cannot be determined from the information given.

17. The mean height of 17 hockey players was 177 centimeters. After the shortest player retired, the mean increased to 178 centimeters. How tall was the player who retired?

    (A) 160 cm
    (B) 161 cm
    (C) 165 cm
    (D) 177 cm
    (E) 178 cm

GO ON TO THE NEXT PAGE

USE THIS SPACE FOR SCRATCHWORK.

18. If $\sin x = \cot x$, which of the following is a possible radian value of x?

(A) -0.618
(B) 0.797
(C) 0.905
(D) 2.113
(E) 2.245

19. For every pair (x, y) in the plane, f reflects (x, y) over the x-axis, and g reflects (x, y) over the origin. Which of the following describes the composite function $g \circ f$ on a given point in the plane?

(A) $g \circ f: (x, y) \rightarrow (-x, -y)$
(B) $g \circ f: (x, y) \rightarrow (x, -y)$
(C) $g \circ f: (x, y) \rightarrow (x, y)$
(D) $g \circ f: (x, y) \rightarrow (-x, y)$
(E) $g \circ f: (x, y) \rightarrow (-2x, -y)$

20. If $\arcsin(\sin x) = 0$, what could be the value of $2x$?

(A) $\frac{\pi}{2}$
(B) π
(C) $\frac{3\pi}{2}$
(D) 2π
(E) 2

21. If the surface area of a sphere increases by 50%, by what percentage does its volume increase?

(A) 22.4%
(B) 35.3%
(C) 50.0%
(D) 83.7%
(E) 100.0%

GO ON TO THE NEXT PAGE

USE THIS SPACE FOR SCRATCHWORK.

22. According to a recent poll, 60% of the population approves of the governor's performance, and 20% of those who approve of her performance are registered in an opposing party. If a member of the population is chosen at random, what is the probability that the person approves of the governor's performance and is registered in an opposing party?

    (A) 8%
    (B) 12%
    (C) 20%
    (D) 48%
    (E) 60%

23. For all θ such that $0 < \theta < \frac{\pi}{2}$, $\frac{\sin\theta\sin(-\theta)}{\cos(-\theta)} =$

    (A) $-\tan\theta$
    (B) $1 - \cos\theta$
    (C) $\cos\theta - \sec\theta$
    (D) $\sec\theta + \cos\theta$
    (E) $1 + \cos\theta$

24. What is the domain of $f(x) = \frac{2a + x}{x - b}$?

    (A) $x > 0$
    (B) $x > b$
    (C) $-b < x < b$
    (D) $x \neq b$
    (E) All real numbers

25. Of the following data sets, which has the largest standard deviation?

    (A) 1, 1, 1, 1, 1
    (B) 1, 2, 3, 4, 5
    (C) 1, 3, 5, 7, 9
    (D) 2, 4, 6, 8, 10
    (E) 2, 4, 8, 16, 32

GO ON TO THE NEXT PAGE

USE THIS SPACE FOR SCRATCHWORK.

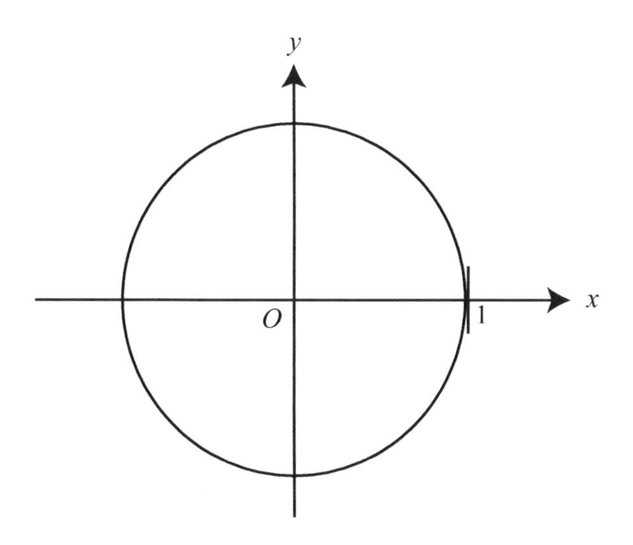

26. The figure above graphs which of the following equations?

(A) $r = 1$
(B) $r = \theta$
(C) $r = \sin\theta + \cos\theta$
(D) $r = \sin^2\theta$
(E) $r = \cos^2\theta$

27. Where defined, $\dfrac{1 - \sin^2\theta}{1 - \cos^2\theta} =$

(A) $\cot\theta$
(B) $1 - \tan^2\theta$
(C) $\tan^2\theta$
(D) $\cot^2\theta$
(E) 1

28. What is the range of the function defined by
$f(x) = \begin{cases} \sqrt[10]{x}, & x > 5 \\ 5x^2, & x \leq 5 \end{cases}$?

(A) $y > 10$
(B) $y > \sqrt[10]{5}$
(C) $y \geq 0$
(D) $y \leq 5$
(E) All real numbers

GO ON TO THE NEXT PAGE

USE THIS SPACE FOR SCRATCHWORK.

29. The graph of the rational function $y = \dfrac{5}{x^2 - 6x + 5}$ has a vertical asymptote at

(A) $x = 5$ only
(B) $x = 1$ only
(C) $x = 0$ only
(D) $x = 1$ and $x = 5$ only
(E) $x = 0, x = 1$ and $x = 5$ only

30. In January, there is a 30.0% chance of snow each day. What is the probability that it will snow each day in the first three days of January?

(A) 2.7%
(B) 9.0%
(C) 10.0%
(D) 30.0%
(E) 90.0%

31. Suppose the graph of $f(x) = -x^3$ is translated 2 units down and 4 units to the right. If the resulting graph represents $g(x)$, what is the y-intercept of g?

(A) -12
(B) 0
(C) 14
(D) 32
(E) 62

32. What is the domain of $f(x) = \sqrt[3]{4 - x^2}$?

(A) $x > 2$
(B) $x > -2$
(C) $-2 < x < 2$
(D) $x \neq 2$ or -2
(E) All real numbers

GO ON TO THE NEXT PAGE

USE THIS SPACE FOR SCRATCHWORK.

33. Two distinct circles, both tangent to the x and y axis, intersect each other at two points. If the center of one circle is given by (x_1, y_1), which of the following must represent the center of the other circle?

 (A) (x_1^2, y_1^2)
 (B) (ax_1, by_1) where $a \neq b$
 (C) (ax_1, y_1)
 (D) (x_1, by_1)
 (E) (cx_1, cy_1) where $c \neq 0$

34. What is the tenth term of the following geometric sequence: 1458, -486, 162, -52...?

 (A) $-28,697,814$
 (B) -0.074
 (C) -0.02
 (D) 0.666
 (E) $9,565,938$

35. If $\log_a a^{x^2} = 6^y$, then $y =$

 (A) 2
 (B) 4
 (C) 6
 (D) 9
 (E) It cannot be determined from the information given.

36. If $f(x,y,z) = (z,x,y)$ and $g(x,y,z) = (y,z,x)$, which of the following is NOT true?

 (A) $g(g(x,y,z)) = f(x,y,z)$
 (B) $f(f(x,y,z)) = g(x,y,z)$
 (C) $f(f(f(x,y,z))) = f(x,y,z)$
 (D) $g(g(g(x,y,z))) = f(f(f(x,y,z)))$
 (E) $f(g(x,y,z) = g(f(x,y,z))$

GO ON TO THE NEXT PAGE

USE THIS SPACE FOR SCRATCHWORK.

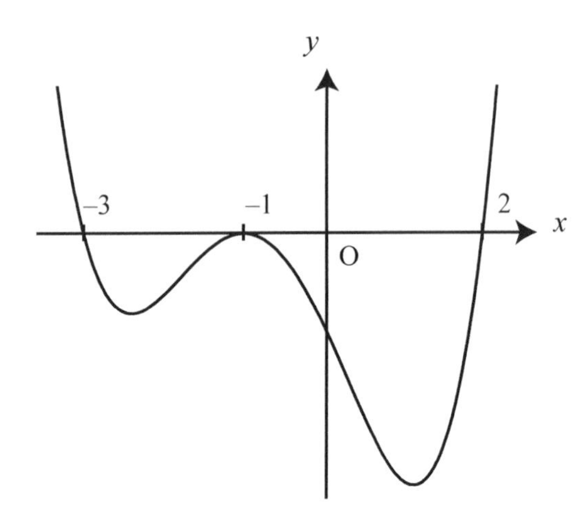

Figure 5

37. The graph of $y = f(x)$ is shown in Figure 5. Which of the following could be $f(x)$?

    (A) $(x-3)(x+1)(x-2)$
    (B) $(x-3)^2(x+1)^2(x-2)$
    (C) $(x+3)(x+1)^2(x-2)$
    (D) $(x+3)^2(x-2)^2$
    (E) $(x-3)(x+1)^2(x-2)$

38. If $0 \leq x \leq \frac{\pi}{2}$ and $\sin x = 0.951$, what is the value of $\tan(2x)$?

    (A) -1.000
    (B) -0.951
    (C) -0.727
    (D) 0.008
    (E) 0.783

GO ON TO THE NEXT PAGE

USE THIS SPACE FOR SCRATCHWORK.

39. Three identical circles are drawn inside a rectangle. They are allowed to intersect with each other but cannot touch the edges of the rectangle. What is the maximum number of distinct regions that could be contained in the rectangle?

(A) 3
(B) 5
(C) 8
(D) 9
(E) 12

40. If $f(x) = \ln(4 + 2x)$, then $f^{-1}(0)=$

(A) -2
(B) $-\frac{3}{2}$
(C) -1
(D) 0
(E) 1

GO ON TO THE NEXT PAGE

USE THIS SPACE FOR SCRATCHWORK.

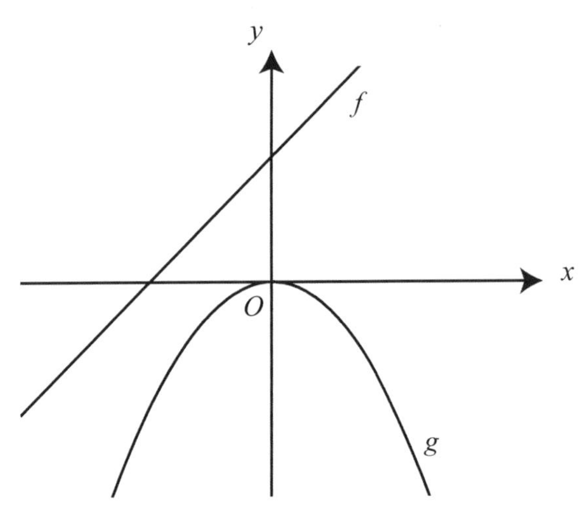

Figure 6

41. Portions of the graphs of f and g are shown in Figure 6. Which of the following could be a portion of the graph of fg?

(A) (B)

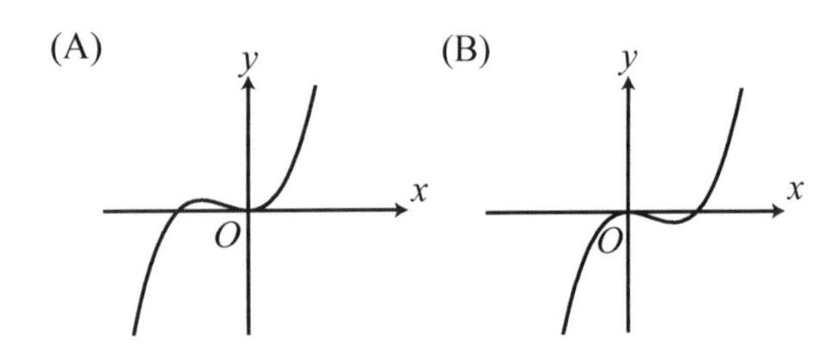

(C) (D)

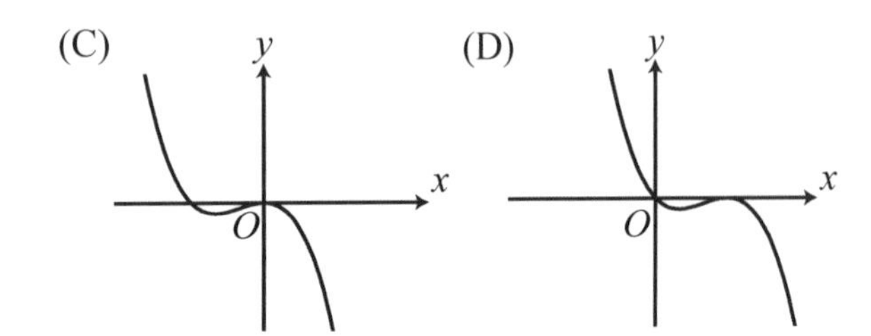

(E)

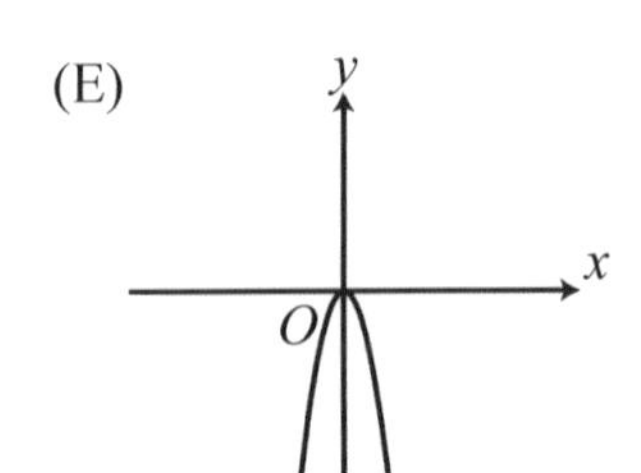

GO ON TO THE NEXT PAGE

USE THIS SPACE FOR SCRATCHWORK.

42. If $x_n = \dfrac{1}{1 + x_{n-1}}$ and $x_0 = 1$, what is the value of x_4?

(A) $\dfrac{3}{5}$

(B) $\dfrac{3}{8}$

(C) $\dfrac{5}{3}$

(D) $\dfrac{5}{8}$

(E) $\dfrac{2}{3}$

43. What is the probability of rolling a fair die twice and getting a prime number both times?

(A) 0.15
(B) 0.25
(C) 0.50
(D) 0.75
(E) 1.00

44. What is the sum of the infinite geometric series $\dfrac{9}{10} + \dfrac{9}{100} + \dfrac{9}{1000} + \dfrac{9}{10000} + \ldots$?

(A) 0.9999
(B) 1

(C) $\dfrac{10}{9}$

(D) 9
(E) 90

GO ON TO THE NEXT PAGE

USE THIS SPACE FOR SCRATCHWORK.

If a figure is a rectangle, it has two parallel sides.

45. Which of the following can be inferred from the statement above?

    (A) If a figure has two parallel sides, it is a rectangle.
    (B) A necessary condition for a figure to have two parallel sides is that it be a rectangle.
    (C) A figure having two parallel sides is implied to be a rectangle.
    (D) If a figure does not have two parallel sides, it is possible that it is a rectangle.
    (E) In order for a figure to be a rectangle, it must have two parallel sides.

46. If $f(x) = x^3 - 3x^2 - x + 3$, which of the following statements are true?

    I. $|f(x)| \geq 3$ for all $x \geq 0$
    II. The function f is increasing for $x \geq 1$
    III. The equation $f(x) = 0$ has no non-real solutions.

    (A) I only
    (B) III only
    (C) I and II only
    (D) I and III only
    (E) II and III only

47. The snack bar sells 50 different snacks. There are 20 different kinds of crackers, 15 different kinds of candy, and 15 different kinds of chips. A coach goes to the snack bar to purchase a variety pack of snacks for his team. How many ways can he put together a variety pack that contains three different crackers, three different candies and four different chips?

    (A) 455
    (B) 1,140
    (C) 1,365
    (D) 1,556,100
    (E) 708,025,500

GO ON TO THE NEXT PAGE

USE THIS SPACE FOR SCRATCHWORK.

48. What is the period of $f(x) = 3\cot(4x - 2)$?

(A) $\frac{\pi}{4}$
(B) $\frac{\pi}{2}$
(C) $\frac{3\pi}{4}$
(D) π
(E) 2π

49. Which of the following is an equation of a line perpendicular to $3x + y = 7$?

(A) $\frac{x}{3} - y = 5$
(B) $\frac{x}{3} + y = 7$
(C) $3x + y = -7$
(D) $\frac{x}{3} + y = -7$
(E) $3x - y = 7$

50. Which of the following states all possible conditions for which $\sqrt[3]{x^2 - y^2}$ is negative?

(A) $x > y$
(B) $y > x$
(C) $|x| > |y|$
(D) $|y| > |x|$
(E) None of the above

STOP

If you finish before time is up, you may check your work on the rest of the test.

SCORING YOUR TEST

A toast

- Congrats on completing your first test.[1]

Overview

- Scoring your test is neither as complicated nor as simple as it could be.[2]
- To score your test, first you'll need to calculate your raw score[3] and then use it to find your scaled score.[4]
- This perhaps should go without saying, but we're not wizards, and we can't predict exactly how you'll do on the real thing.[5]

Getting your raw score

- Compare the answers you entered on your answer sheet with the answer key chart on page 33[6] and mark each answer correct, incorrect, or skipped.[7]
- Use the formula at the bottom of the page to calculate your raw score.
- That is all.

Looking up your scaled score

- Turn the page and we'll tell you how.

1 Well, our first test. Probably your first too, but what do we know?
2 Hooray?
3 That's your total correct minus a quarter of your total incorrect.
4 That's the one you probably really care about. More details on the next page.
5 Think of this as a rough guess.
6 Hint: it's the page that's right next to this one.
7 See how handy the chart is for that? You're welcome.

QUESTION NUMBER	CORRECT ANSWER	YOUR ANSWERS: CORRECT	INCORRECT	SKIP
1	C			
2	D			
3	D			
4	A			
5	C			
6	A			
7	B			
8	A			
9	D			
10	E			
11	D			
12	E			
13	B			
14	C			
15	C			
16	D			
17	B			
18	C			
19	D			
20	D			
21	D			
22	B			
23	C			
24	D			
25	E			

QUESTION NUMBER	CORRECT ANSWER	YOUR ANSWERS: CORRECT	INCORRECT	SKIP
26	A			
27	D			
28	C			
29	D			
30	A			
31	E			
32	E			
33	E			
34	B			
35	E			
36	C			
37	C			
38	C			
39	C			
40	B			
41	C			
42	D			
43	B			
44	B			
45	E			
46	B			
47	E			
48	A			
49	A			
50	D			

TOTAL CORRECT [ ] - .25 × **TOTAL INCORRECT** [ ] = **RAW SCORE** [ ]

Finding your scaled score

- Round your raw score to the nearest digit and look up the corresponding scaled score in the table below.

RAW SCORE	SCALED SCORE
50	800
49	800
48	800
47	800
46	800
45	800
44	800
43	800
42	790
41	780
40	770
39	760
38	750
37	740
36	730
35	720
34	710
33	700
32	690
31	680
30	670

RAW SCORE	SCALED SCORE
29	660
28	650
27	640
26	630
25	630
24	620
23	610
22	600
21	590
20	580
19	570
18	560
17	560
16	550
15	540
14	530
13	530
12	520
11	510
10	500
9	500

RAW SCORE	SCALED SCORE
8	490
7	480
6	480
5	470
4	460
3	450
2	440
1	430
0	410
-1	390
-2	370
-3	360
-4	340
-5	340
-6	330
-7	320
-8	320
-9	320
-10	320
-11	310
-12	310

TEST TWO

SCYLLA, CHARYBDIS, AND FRIENDS

For most accurate results, take this test in a setting similar to the one you'll take the real test in. We've organized the essentials into this simple mnemonic:

Locate yourself somewhere quiet and free from distractions.

Let a scientific or graphing calculator be your buddy.

Answer questions on an answer sheet in the back of this book or downloaded from barebonesguide.com/answersheet.

Make sure your workspace is free of outside references.

A timer near you should be set to one hour.

After you have finished, consult our scoring instructions at the end of the test to calculate your raw and scaled scores.

MATHEMATICS LEVEL 2 TEST

REFERENCE INFORMATION

YOU MAY USE THE FOLLOWING INFORMATION AS YOU TAKE THIS TEST.

A right circular cone with radius r and height h has volume $V = \frac{1}{3}\pi r^2 h$

A right circular cone with circumference of the base c and slant height ℓ has lateral area $S = \frac{1}{2}c\ell$

A sphere with radius r has volume $V = \frac{4}{3}\pi r^3$

A sphere with radius r has surface area $S = 4\pi r^2$

A pyramid with base area B and height h has volume $V = \frac{1}{3}Bh$

GO ON TO THE NEXT PAGE

MATHEMATICS LEVEL 2 TEST

Choose the best choice given for each of the following questions. The exact value may not be one of the choices; in this case select the best approximation of the true value.

Note the following items carefully.

(1) You will need a scientific or graphing calculator to answer some of the questions in this test. Some of these questions will require you to determine whether your calculator should be in degree or radian mode. For other questions, however, you will not need a calculator at all.

(2) All figures are drawn to scale and lie in the plane unless otherwise specified.

(3) For the purposes of this test, the domain of any function f is the set of real numbers x for which $f(x)$ is a real number and the range of f is the set of real numbers $f(x)$, where x is in the domain of f. Any deviation from these assumptions will be noted in each problem to which it applies.

USE THIS SPACE FOR SCRATCHWORK.

1. What is the value of x^2 if $x = \sqrt{20^2 - 13^2}$?

(A) $\sqrt{7}$
(B) $\sqrt{231}$
(C) 49
(D) 231
(E) 53,361

GO ON TO THE NEXT PAGE

USE THIS SPACE FOR SCRATCHWORK.

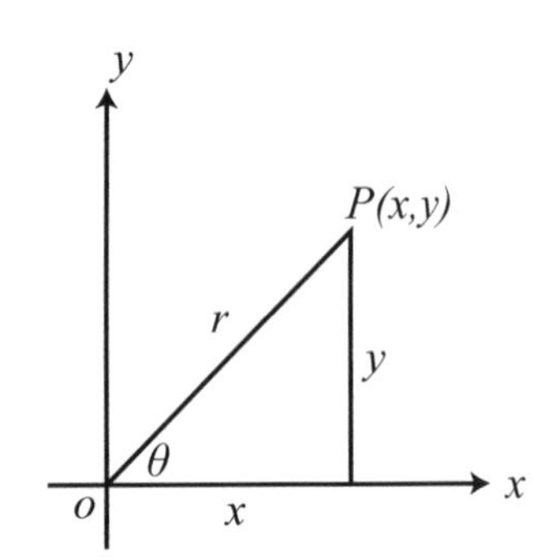

Figure 1

2. In Figure 1, $x \tan\theta =$

(A) x
(B) y
(C) r
(D) $\frac{x^2}{y}$
(E) $\frac{y^2}{x}$

3. Which of the following is an equation of a line parallel to $4x + 2y = 8$?

(A) $2x - y = 6$
(B) $2x + 2y = 3$
(C) $2x - 4y = 8$
(D) $4x - 2y = 8$
(E) $2x + y = 7$

GO ON TO THE NEXT PAGE

USE THIS SPACE FOR SCRATCHWORK.

4. If $\cos\theta = \tan\theta$, θ could be in which quadrant?

   (A) I only
   (B) III only
   (C) I and II only
   (D) I and IV only
   (E) III and IV only

5. If $0 \leq x < 2\pi$ and $\sin x = -0.85$, $x =$

   (A) 0.85
   (B) 1.57
   (C) 2.13
   (D) 3.14
   (E) 5.27

6. The only prime factors of a number n are 3, 7, 11, and 13. Which of the following CANNOT be a factor of n?

   (A) 33
   (B) 73
   (C) 91
   (D) 121
   (E) 189

7. On the interval $0 \leq \theta < \pi$, if $\tan\theta = 1.9626$, then $\sin\theta =$

   (A) -1.100
   (B) -0.891
   (C) 0.510
   (D) 0.891
   (E) 1.100

GO ON TO THE NEXT PAGE

USE THIS SPACE FOR SCRATCHWORK.

8. There are three types of candies in a jar: Xcites, Yummies, and Zig-zags. There are half as many Yummies as Xcites. There are five times as many Zig-zags as Yummies. If there are a total of 16 candies in the jar, how many Zig-zags are there?

   (A) 2
   (B) 4
   (C) 8
   (D) 10
   (E) 12

9. If $f(x) = 8(x - 5)$ and $g(x) = x^3 + 10$, which of the following expresses $f(g(a))$?

   (A) $8(a^3 + 5)$
   (B) $8(a^4 - 5) + 10$
   (C) $8(a^3 - 5) + 10$
   (D) $a^4 - 15$
   (E) $(8a - 5)^3 + 10$

10. The history club is planning a trip for club members and their chaperones to Medieval Times, which has an admission price of $11 per person. Admission for 5 chaperones and the $300 cost of the bus must be split evenly among the club members. Which of the following expresses the cost c in dollars that each club member much contribute as a function of n, the number of members on the trip?

    (A) $c(n) = \dfrac{300 + 11n}{n}$
    (B) $c(n) = \dfrac{355 + 11n}{n - 2}$
    (C) $c(n) = \dfrac{300 + 11n}{n + 2}$
    (D) $c(n) = \dfrac{300 + 11n}{n - 2}$
    (E) $c(n) = \dfrac{355 + 11n}{n}$

GO ON TO THE NEXT PAGE

$$a+b=3$$
$$b+c=7$$
$$a+b+c=11$$

11. What is the value of b?

(A) -1
(B) 0
(C) $\frac{7}{8}$
(D) 1
(E) $\frac{8}{7}$

12. If $f(x)=|7-2x^2|$, then $f(3)=$

(A) $f(-\sqrt{2})$
(B) $f(-2)$
(C) $f(2)$
(D) $f(1)$
(E) $f(-3)$

13. What is $\arccos\left(\cos\left(\frac{-\pi}{3}\right)\right)$?

(A) 0
(B) $\frac{\pi}{6}$
(C) $\frac{\pi}{3}$
(D) $\frac{1}{2}$
(E) 1

14. For all θ, $\sin\theta-\sin(-\theta)+\tan(-\theta)=$

(A) $\tan\theta$
(B) $-\tan\theta$
(C) $-\sin\theta+\tan\theta$
(D) $2\sin\theta+\tan\theta$
(E) $2\sin\theta-\tan\theta$

GO ON TO THE NEXT PAGE

USE THIS SPACE FOR SCRATCHWORK.

15. The graph of the rational function f, where $f(x) = \dfrac{x^2 - 3x}{x^3 - x^2 - 6x}$, has a vertical asymptote at

    I. $x = -2$
    II. $x = 3$
    III. $x = 0$

    (A) I only
    (B) II only
    (C) I and III only
    (D) II and III only
    (E) I, II, and III

16. If $f(3x + 7) = 6x + 1$ for all real numbers x, then $f(x) =$

    (A) $2x - 13$
    (B) $18x + 10$
    (C) $18x + 42$
    (D) $\dfrac{1}{2}x + \dfrac{13}{2}$
    (E) $6x + 7$

GO ON TO THE NEXT PAGE

USE THIS SPACE FOR SCRATCHWORK.

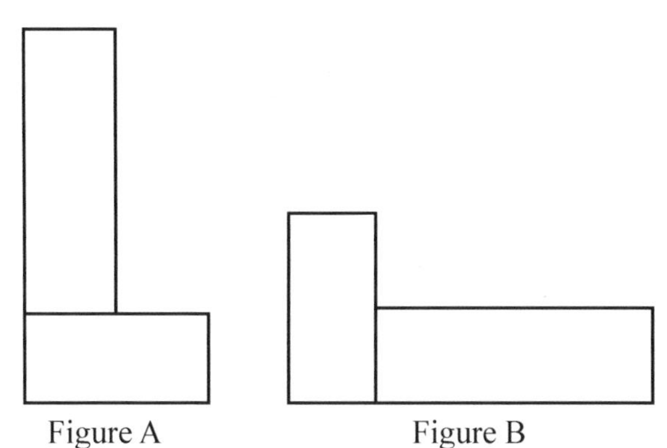
Figure A Figure B

17. Which of the following transformations corresponds to the change from Figure A to Figure B?

    (A) Clockwise rotation by 90°
    (B) Clockwise rotation by 90° followed by reflection over the vertical axis
    (C) Reflection over the horizontal axis followed by reflection over the vertical axis
    (D) Counterclockwise rotation by 90° followed by reflection over the horizonal axis
    (E) Counterclockwise rotation by 90° followed by reflection over the vertical axis

18. If $\sqrt{x+9} = x - 3$, what is the complete solution set for x?

    (A) $\{-7, 0, 7\}$
    (B) $\{-3, 0\}$
    (C) $\{0, 7\}$
    (D) $\{3, 7\}$
    (E) $\{7\}$

GO ON TO THE NEXT PAGE

USE THIS SPACE FOR SCRATCHWORK.

$$y^2 + 4y = 2x - x^2 - 4$$

19. What is the length of the radius of the circle determined by the equation given above?

(A) 1
(B) 3
(C) $\sqrt{11}$
(D) 4
(E) 16

20. What is the range of the function $y = 3\sin\left(3x + \frac{\pi}{4}\right)$?

(A) $[-3]$
(B) $(-3, 3)$
(C) $[0, \frac{\pi}{4}]$
(D) $[-3, 3]$
(E) $(-\infty, \infty)$

21. For which function does $f(-x) = f(x)$?

(A) $f(x) = \sin x$
(B) $f(x) = \sin^2 x$
(C) $f(x) = x^2 \sin x$
(D) $f(x) = \sin(x - \pi)$
(E) $f(x) = x^3$

22. If $f(x) = \frac{x^3 - 5}{6}$, what is $f^{-1}(\frac{1}{2})$?

(A) -1.26
(B) 0.81
(C) 1.72
(D) 2
(E) 3

GO ON TO THE NEXT PAGE

USE THIS SPACE FOR SCRATCHWORK.

23. If $4^x = 32^{\frac{1}{y}}$, what is the value of xy?

(A) 0.33
(B) 1.79
(C) 2.50
(D) 3.00
(E) 19.93

24. Last week, 2,504,705 residents of Kings County and 1,585,870 residents of New York County voted on a proposition. The percentage of people against the proposition was the same in each county, and a total of 818,115 residents from both counties voted against the proposition. How many Kings County residents voted against the proposition?

(A) 317,174
(B) 500,941
(C) 510,648
(D) 723,692
(E) 1,073,923

25. Which of the following is equivalent to xy^3 for x and y not equal to zero?

(A) $\frac{x^4y}{y^{-2}x^3}$

(B) $\frac{x^4y^{-2}}{yx^3}$

(C) $\frac{x^3y^4}{y^{-2}x^4}$

(D) $\frac{x^4y}{y^3x^{-2}}$

(E) $\frac{xy^4}{y^{-2}x^3}$

GO ON TO THE NEXT PAGE

USE THIS SPACE FOR SCRATCHWORK.

26. Compared to the graph of $y = -2\cos 3x$, what is the shift in the graph of the function $y = -2\cos 3\left(x - \frac{\pi}{6}\right)$?

    (A) 2 units to the left
    (B) $\frac{\pi}{6}$ units to the left
    (C) $\frac{\pi}{6}$ units to the right
    (D) $\frac{\pi}{2}$ units to the right
    (E) $\frac{\pi}{2}$ units to the left

27. What is the area of the triangle whose vertices lie at the x- and y-intercepts of the parabola $y = -2x^2 + 4x + 6$?

    (A) 3
    (B) 4
    (C) 6
    (D) 12
    (E) 24

28. Which of the following functions has an inverse whose domain is all real numbers?

    (A) $\sqrt{y} = x$
    (B) $y^3 = x$
    (C) $y = (x + 1)(x - 1)$
    (D) $y = \frac{1}{x}$
    (E) $y = x^2$

GO ON TO THE NEXT PAGE

USE THIS SPACE FOR SCRATCHWORK.

29. Let $A = (x, y), B = (-x, -y), C = (x, -y)$, and let A' be the midpoint of AB, B' be the midpoint of BC, and C' be the midpoint of CA. What is the area of triangle $A'B'C'$?

(A) $\frac{xy}{2}$

(B) $2xy$

(C) xy

(D) $\frac{x^2}{2}$

(E) $\frac{y^2}{2}$

30. For all θ such that $\frac{\pi}{2} < \theta \leq \pi$ and $\sin\left(\theta + \frac{\pi}{6}\right) = .54$, what is $\cos\theta$?

(A) -0.999
(B) -0.459
(C) 0.047
(D) 0.842
(E) 0.999

s	t
2	8
4	13
6	20
8	24
10	32

31. Which of the following equations most closely fits the data in the table above?

(A) $t = s + 6$
(B) $t = 4s$
(C) $t = 3s + 2$
(D) $t = 6s - 10$
(E) $t = s^2 - 5$

GO ON TO THE NEXT PAGE

USE THIS SPACE FOR SCRATCHWORK.

32. A jet takes off at an angle of elevation of 23° with a speed of 172 m/s along its trajectory. Assuming its speed and direction remain constant, what is its altitude in meters after 5 seconds?

    (A) 67.21
    (B) 336.03
    (C) 365.05
    (D) 791.63
    (E) 2, 201.00

33. A set S contains 12 different positive integers less than 30. If the largest one is removed, which of the following CANNOT be the median of the remaining 11 integers in S?

    (A) 6
    (B) 7
    (C) 8
    (D) 23
    (E) 24

$$f(x) = \frac{4x^3 - 3x^2 + 8}{2x^3 + 3x - 4}$$

34. Approximately what is $f(350, 498)$?

    (A) 0
    (B) 1
    (C) 2
    (D) 350,500
    (E) 701,000

35. The balance of a certain retirement account at any time t can be calculated by the function $B(t) = Pe^{0.08t}$, where P is the initial balance, and t is the elapsed time in years. Approximately how many years will it take for an initial balance of \$1, 000 to grow to \$100, 000?

    (A) 10 years
    (B) 18 years
    (C) 47 years
    (D) 58 years
    (E) 65 years

GO ON TO THE NEXT PAGE

USE THIS SPACE FOR SCRATCHWORK.

36. Each time Kyle parks his car, there is a $\frac{1}{4}$ probability that he hits a nearby car. If Kyle parks his car 4 times today, what is the probability that he hits exactly 2 cars?

    (A) .035
    (B) .211
    (C) .500
    (D) .625
    (E) .844

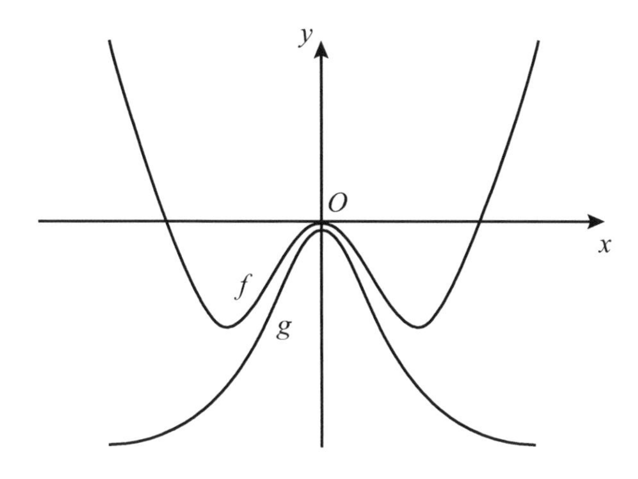

37. The graphs of functions f and g are shown above. Which of the following functions is positive everywhere?

    (A) $f + g$

    (B) $f - g$

    (C) fg

    (D) $\frac{f}{g}$

    (E) fg^2

GO ON TO THE NEXT PAGE

USE THIS SPACE FOR SCRATCHWORK.

38. If $f: (x, y) \to (2x, y + x)$ for every pair (x, y) in the xy-coordinate plane, for what points (x, y) is it true that $(x, y) \to (x, y)$?

(A) The set of points (x, y) such that $y = x$
(B) The set of points (x, y) such that $x = \frac{1}{2}$
(C) The set of points (x, y) such that $y = 0$
(D) The set of points (x, y) such that $x = 0$
(E) $(0, 0)$ only

If x is a whole number, it is an integer.

39. If x is a real number, which of the following CANNOT be logically inferred from the above statement?

I. In order for x to be an integer, it is necessary for x to be a whole number.
II. If x is not a whole number, then x is not an integer.
III. If x is not an integer, then x is not a whole number.

(A) I only
(B) III only
(C) I and II only
(D) II and III only
(E) I, II, and III

40. The intersection of perpendicular planes yields which of the following?

(A) A line
(B) A plane
(C) A point
(D) A sphere
(E) None of the above

GO ON TO THE NEXT PAGE

USE THIS SPACE FOR SCRATCHWORK.

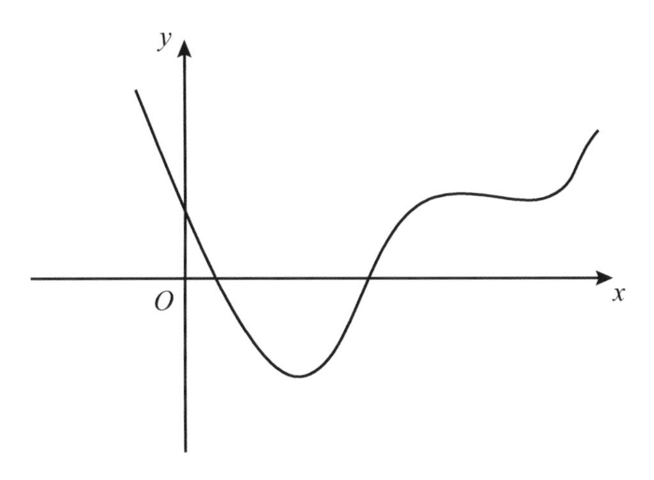

41. The graph of $f(x)$ is given above. If a is some positive number, which of the following graphs CANNOT pass through the origin?

    (A) $f(x + a)$
    (B) $f(x) + a$
    (C) $f(x) - a$
    (D) $a - f(x)$
    (E) $axf(x)$

$$ax^7 + bx^4 - cx + d = 0$$

42. If a, b, c, and d are integers, what is the maximum number of non-real roots for the above equation?

    (A) 7
    (B) 6
    (C) 5
    (D) 4
    (E) 2

GO ON TO THE NEXT PAGE

USE THIS SPACE FOR SCRATCHWORK.

$$f(x) = \frac{2x}{a - x}$$

43. For what value of a does $\lim\limits_{x \to 3} = 1$?

(A) 9
(B) 6
(C) 3
(D) 0
(E) -3

44. In what quadrant of the complex plane is the point uv if $u = (3 + i)$ and $v = (-1 - 2i)$?

(A) I
(B) II
(C) III
(D) IV
(E) It lies on the y-axis.

$$f(x) = \log_b x$$
$$g(x) = \log_b (bx^2)$$
$$h(x) = f(x) + g(x)$$

45. Which of the following is equal to $h(x)$?

(A) $3\log_b x + 1$
(B) $3b\log_b x + 1$
(C) $3\log_b x$
(D) $\log_b 2bx$
(E) $\log_b (bx^2 + x)$

GO ON TO THE NEXT PAGE

USE THIS SPACE FOR SCRATCHWORK.

46. Which of the following must be a solution to the equation $ax^4 + bx^2 + c = 0$?

(A) $\sqrt{\dfrac{-b - \sqrt{b^2 - 4ac}}{2a}}$

(B) $\dfrac{-b + \sqrt{b^2 - 4ac}}{2a}$

(C) $\dfrac{b + \sqrt{b^2 - 4ac}}{2a}$

(D) $\sqrt{\dfrac{-c}{b + a}}$

(E) $\left(\dfrac{-b - \sqrt{b^2 - 4ac}}{2a}\right)^2$

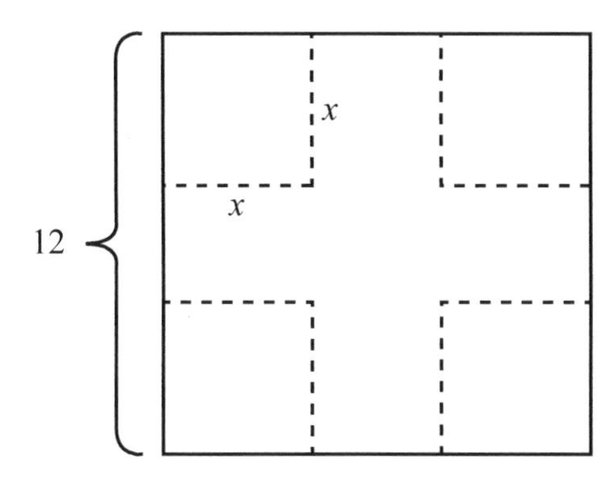

47. A square sheet with side length 12 has squares of side length x cut from the corners, as shown in Figure 4. The resulting shape is folded to make an open box of maximum possible volume. What is x?

(A) 6
(B) 3
(C) 2
(D) $\dfrac{5}{3}$
(E) 1

USE THIS SPACE FOR SCRATCHWORK.

48. A cylinder of height 4 and radius 3 is cut in half into two cylinders of equal volume. What is the lateral surface area of one of the halves?

    (A) 16π
    (B) 12π
    (C) 9π
    (D) 8π
    (E) 6π

49. Which of the following could be the graph of the polar equation $r = 2\theta$?

(A)
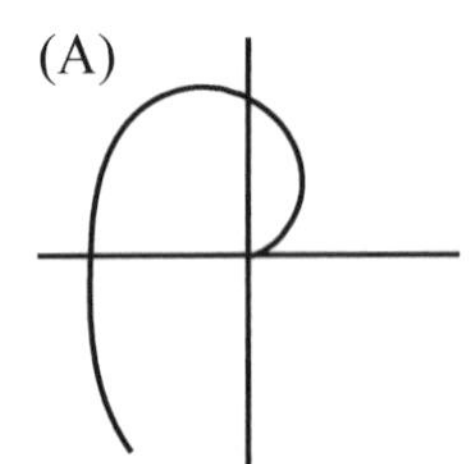

(B)
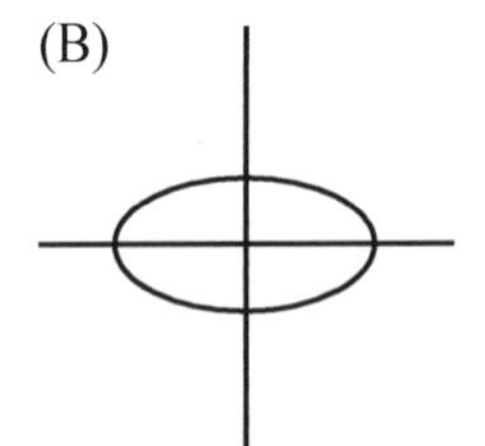

(C)
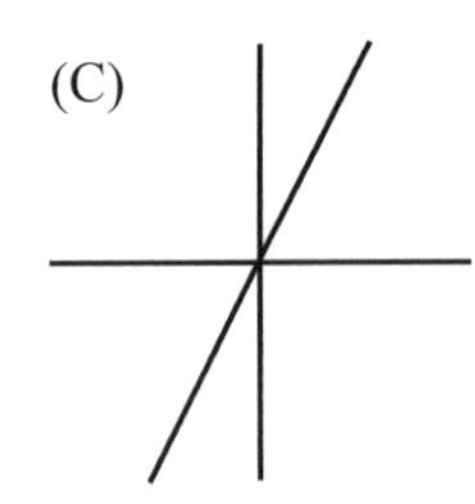

(D)
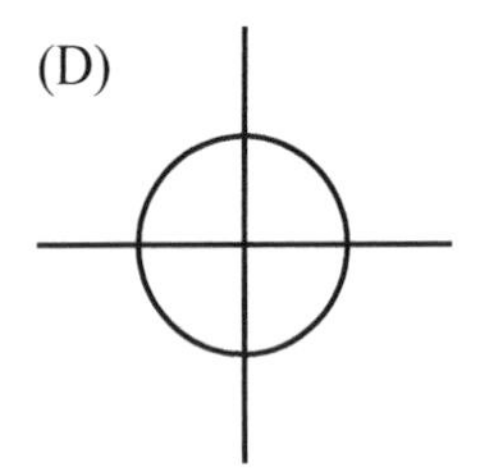

(E)
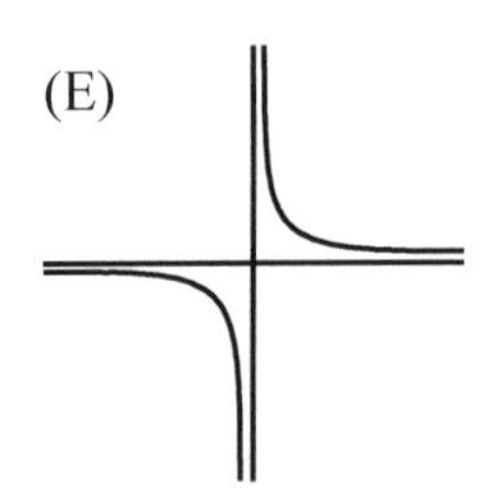

GO ON TO THE NEXT PAGE

USE THIS SPACE FOR SCRATCHWORK.

50. If $-1 < x < 0$, what is the sum of $1 + x + x^2 + x^3 + \ldots$?

(A) $\frac{1}{1-x}$

(B) $\frac{1}{x}$

(C) $2x$

(D) $\frac{1}{x+1}$

(E) $1 + 10x$

STOP

If you finish before time is up, you may check your work on the rest of the test.

SCORING YOUR TEST

Another toast

- Congrats on completing our second test.[1]

Overview

- You'll want to calculate your raw score[2] and then use it to look up your scaled score.
- We do have a crystal ball, but crystal balls don't actually predict the future—your score on the real test may differ from your score on this test.[3]

Getting your raw score

- Compare the answers you entered on your answer sheet with the answer key chart on the facing page and mark each answer correct, incorrect, or skipped.[4]
- Use the formula at the bottom of the page to calculate your raw score.

Looking up your scaled score

- Next page, please.

---

1 And perhaps your second test too.
2 Well, *want* is kind of a strong word, but it's advisable to do it anyway.
3 And anyone who says different is giving you the business.
4 This chart just gets better with repetition, don't you think?

QUESTION NUMBER	CORRECT ANSWER	YOUR ANSWERS: CORRECT	INCORRECT	SKIP
1	D			
2	B			
3	E			
4	C			
5	E			
6	B			
7	D			
8	D			
9	A			
10	E			
11	A			
12	E			
13	C			
14	E			
15	A			
16	A			
17	E			
18	E			
19	A			
20	D			
21	B			
22	D			
23	C			
24	B			
25	A			

QUESTION NUMBER	CORRECT ANSWER	YOUR ANSWERS: CORRECT	INCORRECT	SKIP
26	C			
27	D			
28	B			
29	A			
30	B			
31	C			
32	B			
33	E			
34	C			
35	D			
36	B			
37	B			
38	D			
39	C			
40	A			
41	B			
42	B			
43	A			
44	C			
45	A			
46	A			
47	C			
48	B			
49	A			
50	A			

TOTAL CORRECT		TOTAL INCORRECT		RAW SCORE
	- .25 ×		=	

Finding your scaled score

- Round your raw score to the nearest digit.
- Look up the corresponding scaled score in the table below.
- Laugh a triumphant laugh.[1]

RAW SCORE	SCALED SCORE
50	800
49	800
48	800
47	800
46	800
45	800
44	800
43	800
42	790
41	780
40	770
39	760
38	750
37	740
36	730
35	720
34	710
33	700
32	690
31	680
30	670

RAW SCORE	SCALED SCORE
29	660
28	650
27	640
26	630
25	630
24	620
23	610
22	600
21	590
20	580
19	570
18	560
17	560
16	550
15	540
14	530
13	530
12	520
11	510
10	500
9	500

RAW SCORE	SCALED SCORE
8	490
7	480
6	480
5	470
4	460
3	450
2	440
1	430
0	410
-1	390
-2	370
-3	360
-4	340
-5	340
-6	330
-7	320
-8	320
-9	320
-10	320
-11	310
-12	310

1 This is generally a good idea regardless of your current level of triumph.

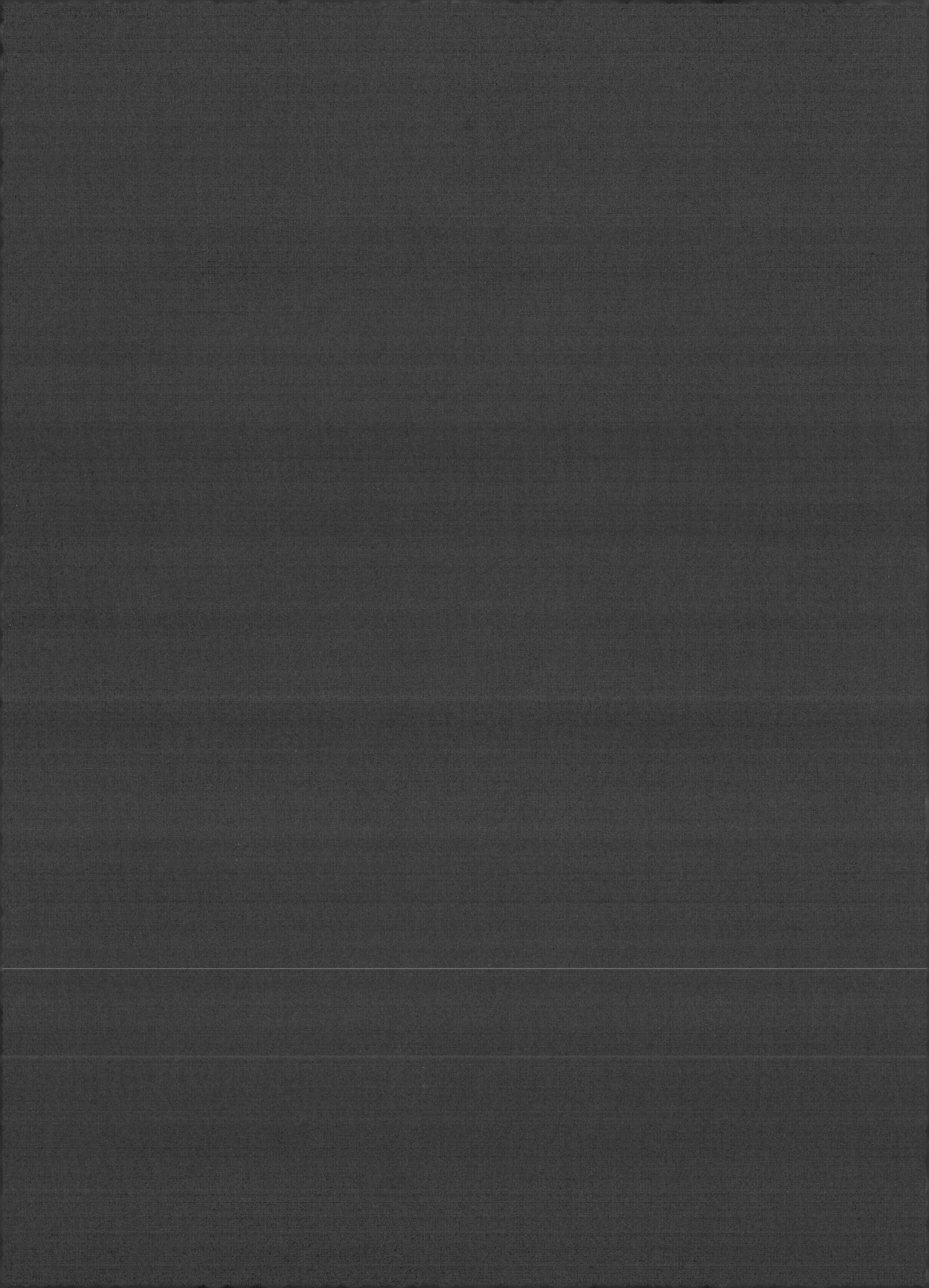

TEST THREE
STANDARDIZED TASTE

For most accurate results, take this test in a setting similar to the one you'll take the real test in. We've organized the essentials into this simple mnemonic:

Select a quiet spot free from distractions.

Place a timer within sight (and set it to one hour).

Other than your test, calculator, bubble sheet, and pencils, your workspace should be clear—no outside references.

Record your answers on an answer sheet in the back of this book or downloaded from barebonesguide.com/answersheet.

Even though it isn't technically required, having a scientific or graphing calculator is pretty much a necessity for this test.

After you have finished, consult our scoring instructions at the end of the test to calculate your raw and scaled scores.

MATHEMATICS LEVEL 2 TEST

REFERENCE INFORMATION

YOU MAY USE THE FOLLOWING INFORMATION AS YOU TAKE THIS TEST.

A right circular cone with radius r and height h has volume $V = \frac{1}{3}\pi r^2 h$

A right circular cone with circumference of the base c and slant height ℓ has lateral area $S = \frac{1}{2}c\ell$

A sphere with radius r has volume $V = \frac{4}{3}\pi r^3$

A sphere with radius r has surface area $S = 4\pi r^2$

A pyramid with base area B and height h has volume $V = \frac{1}{3}Bh$

GO ON TO THE NEXT PAGE

MATHEMATICS LEVEL 2 TEST

Choose the best choice given for each of the following questions. The exact value may not be one of the choices; in this case select the best approximation of the true value.

Note the following items carefully.

(1) You will need a scientific or graphing calculator to answer some of the questions in this test. Some of these questions will require you to determine whether your calculator should be in degree or radian mode. For other questions, however, you will not need a calculator at all.

(2) All figures are drawn to scale and lie in the plane unless otherwise specified.

(3) For the purposes of this test, the domain of any function f is the set of real numbers x for which $f(x)$ is a real number and the range of f is the set of real numbers $f(x)$, where x is in the domain of f. Any deviation from these assumptions will be noted in each problem to which it applies.

USE THIS SPACE FOR SCRATCHWORK.

1. $\sin\theta - \sin(\pi - \theta) =$

(A) 0
(B) 1
(C) $2\sin\theta$
(D) $-\cos\theta$
(E) $\cos\theta\sin\theta$

2. Which of the following is a zero of the polynomial $f(x) = 2x^2 + 3x - 20$?

(A) -4
(B) -2.5
(C) -2
(D) 0
(E) .5

GO ON TO THE NEXT PAGE

USE THIS SPACE FOR SCRATCHWORK.

3. In how many places does a circle centered at (12, 6) with radius 8 touch the x- and y-axes?

(A) 0
(B) 1
(C) 2
(D) 3
(E) 4

4. If $e^{2x} = 6$, $x =$

(A) 0.90
(B) 1.79
(C) 3
(D) 5.44
(E) 16.31

5. If $f(3x - 2) = 3x + 3$ for all real numbers x, then $f(x) =$

(A) $x - 1$
(B) $3x + 1$
(C) $x + 1$
(D) $x + 5$
(E) $3x - 3$

6. In what quadrant(s) do the graphs of the equations $y = 2x + 4$ and $y = 3x^2 - 4x + 1$ intersect?

(A) I only
(B) II only
(C) III only
(D) I and II only
(E) II and III only

GO ON TO THE NEXT PAGE

USE THIS SPACE FOR SCRATCHWORK.

7. After a 20% raise, Jeffrey's salary is \$50,000, and after a 30% raise, Julie's salary is \$52,000. What was the approximate difference in salaries before the raises?

   (A) \$0
   (B) \$1,667
   (C) \$2,000
   (D) \$3,133
   (E) \$3,333

8. A man on top of a 24-foot building looks down at an angle of 24° below the horizontal and sees a dog. How far from the base of the building is the dog?

   (A) $\dfrac{24}{\tan 24^\circ}$
   (B) $\dfrac{24}{\sin 24^\circ}$
   (C) $\dfrac{\sin 24^\circ}{24}$
   (D) $24\tan 24^\circ$
   (E) $24\cos 24^\circ$

9. If x, y and z are nonzero real numbers, $x^9y^{-3}z^5 = 7x^8y^{-2}z^4$, then $x =$

   (A) $\dfrac{7}{y^5}$
   (B) $\dfrac{7y}{z}$
   (C) $7x$
   (D) $\dfrac{7}{yz}$
   (E) $7y^5z^9$

10. If $2x^2 + 5 = 37$ and $4y - 4 = 8$, what is the least possible value for $x + y$?

    (A) -12
    (B) -1
    (C) 3
    (D) 4
    (E) 7

GO ON TO THE NEXT PAGE

USE THIS SPACE FOR SCRATCHWORK.

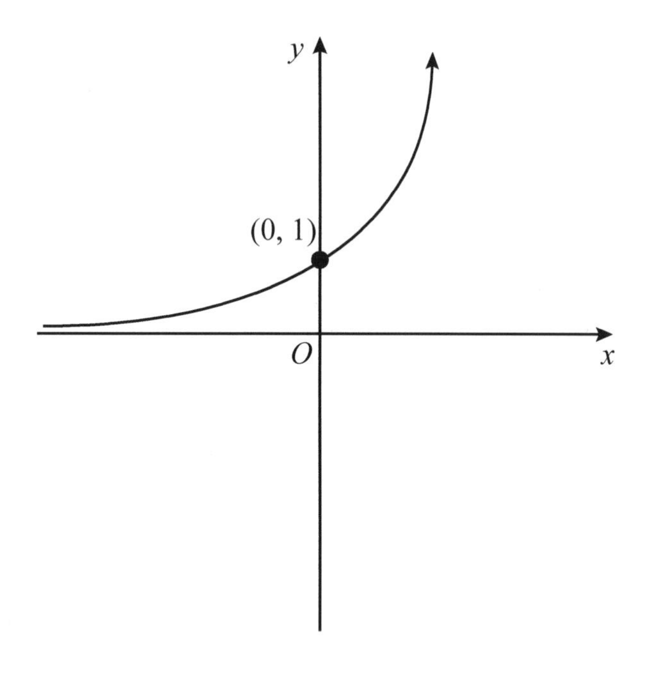

11. The figure above shows the graph of $f(x)$. If $g(f(x)) = x$, which of the following could be the graph of $g(x)$?

(A)
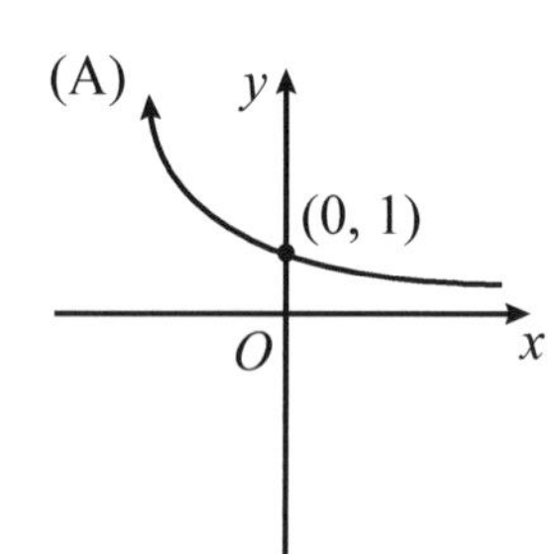

(B)
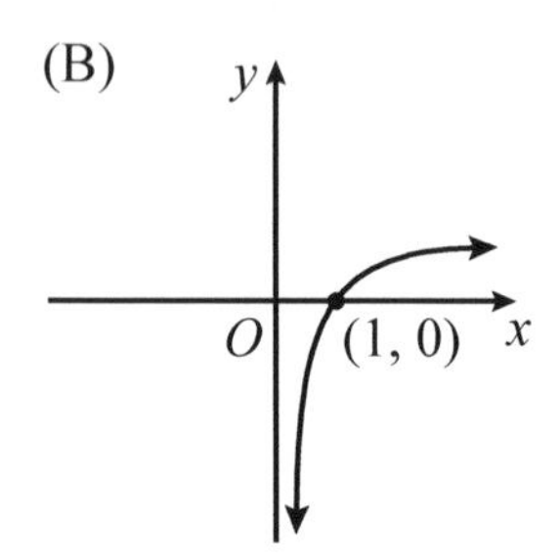

(C)
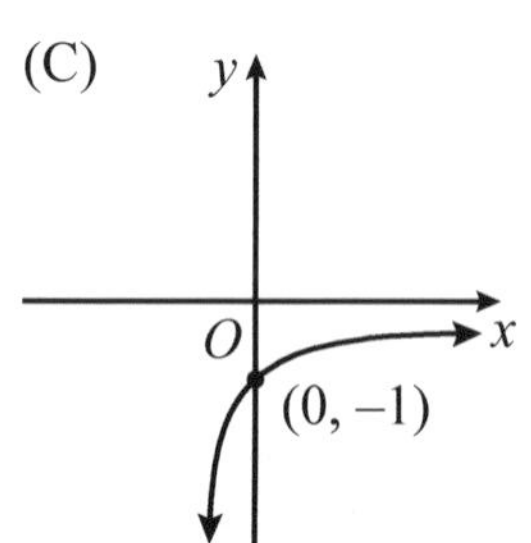

(D)
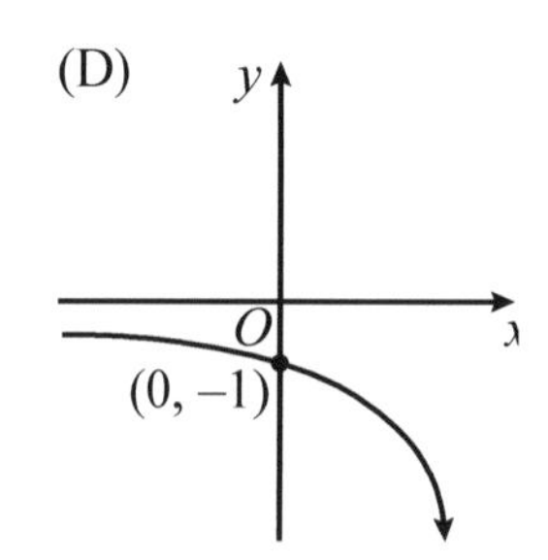

(E)
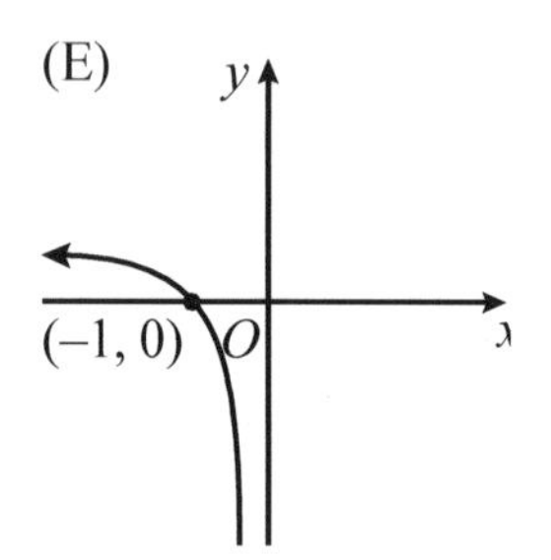

GO ON TO THE NEXT PAGE

USE THIS SPACE FOR SCRATCHWORK.

12. A triangle is wholly contained in the first quadrant and its vertex A has coordinates (x, y). If the triangle were rotated 180° about the origin, what would be the coordinates of the rotated vertex, A^1?

    (A) (x, y)
    (B) $(-x, -y)$
    (C) (y, x)
    (D) $(-y, -x)$
    (E) $(-x, y)$

13. What is the minimum value of the function $y = 2\sin x - 5$ for $0 \leq x < 2\pi$?

    (A) -7
    (B) -5
    (C) -2
    (D) 2
    (E) 5

14. The average length of an article in a newspaper that publishes 50 articles is 750 words. If an additional article decreases the average to 740 words, how long is the additional article?

    (A) 10 words
    (B) 200 words
    (C) 240 words
    (D) 250 words
    (E) 740 words

15. The growth rate r of the squirrel population in Central Park is modeled by the equation $r(x) = x(A - x)$, where x is the population and A is a positive integer. At which population level is the growth rate the greatest?

    (A) 1
    (B) $\frac{A}{2}$
    (C) A
    (D) $\frac{3A}{2}$
    (E) $2A$

GO ON TO THE NEXT PAGE

USE THIS SPACE FOR SCRATCHWORK.

16. The intersection of a plane and a cone could result in which of the following?

I. A point
II. A circle
III. An ellipse

(A) I only
(B) I and II only
(C) I and III only
(D) II and III only
(E) I, II, and III

17. How many different ways can Puppy Bowl organizers pick a team of 5 basset hounds, 3 Chihuahuas, and 3 miniature poodles from a pool of 10 basset hounds, 8 Chihuahuas, and 5 miniature poodles?

(A) 120
(B) 400
(C) 4,320
(D) 141,120
(E) 609,638,400

18. If $f(g(x)) = \dfrac{3(x^2 - 2x + 1) + 1}{x - 1}$ and $f(x) = \dfrac{3x^2 + 1}{x}$, then $g(x) =$

(A) $x - 2$
(B) $x - 1$
(C) $x^2 - 2x + 1$
(D) $x^2 + 1$
(E) $x^2 + 2$

GO ON TO THE NEXT PAGE

USE THIS SPACE FOR SCRATCHWORK.

19. If p is a positive integer, and 11 is not a factor of p, which of the following CANNOT be true?

(A) 11 is a factor of $9p + 2p$.

(B) 11 is a factor of $8p^2 + 3p$.

(C) p is a factor of 11.

(D) 11 is a factor of p^{11}.

(E) 11 is a factor of $4p^5 + 7p^6$.

20. What is the domain of $f(x) = \sqrt[4]{-x - 3}$?

(A) $x < -3$
(B) $x > -3$
(C) $x \geq -3$
(D) $x \leq -3$
(E) All real numbers

If $x = -5$, then $\sqrt{x}$ is not a real number.

21. An indirect proof of the above statement could begin with the assumption that

(A) $x < 0$
(B) $x = -5$
(C) $\sqrt{x}$ is positive
(D) $\sqrt{x}$ is a real number
(E) $x^2 = 25$

22. For all non-zero numbers x and y, $x \otimes y = \dfrac{x^2 + 2xy + y^2}{2xy}$.

If $4 \otimes n = \dfrac{-1}{4}$, what is a possible value of n?

(A) -6
(B) -2
(C) 2
(D) 4
(E) 8

GO ON TO THE NEXT PAGE

USE THIS SPACE FOR SCRATCHWORK.

23. $\dfrac{n(n-2)!}{(n-1)!(n+1)} =$

(A) $\dfrac{n}{(n-1)^2}$

(B) $\dfrac{n-2}{n+1}$

(C) $\dfrac{n}{n^2-1}$

(D) $\dfrac{n(n-1)}{(n-1)(n+1)}$

(E) $\dfrac{n(n-2)}{n^2-1}$

24. If $f(x) = ax^2 + bx + c$ for all real numbers x and if $f(0) = 5$ and $f(1) = 4$, then $a + b =$

(A) -9
(B) -5
(C) -1
(D) 0
(E) 4

25. If x, y, and z are positive integers with $x + y \leq 10$ and $y + z \geq 15$, what is the least possible value for z?

(A) 1
(B) 3
(C) 5
(D) 6
(E) 7

GO ON TO THE NEXT PAGE

USE THIS SPACE FOR SCRATCHWORK.

26. Which of the following diagrams could represent a piece of cardboard that can be folded to make a closed-top box with volume $4x^3$?

(A)

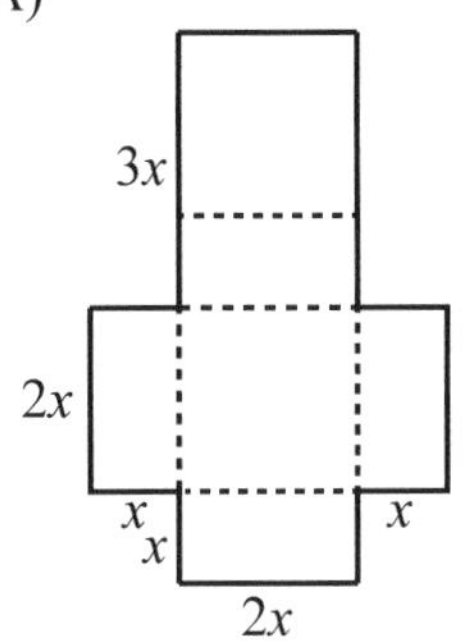

(B)

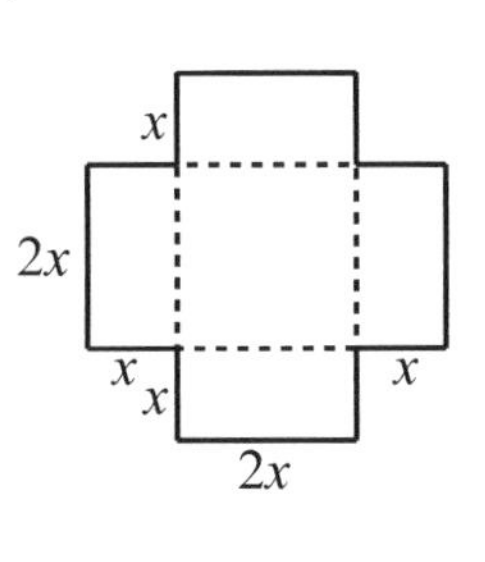

(C)

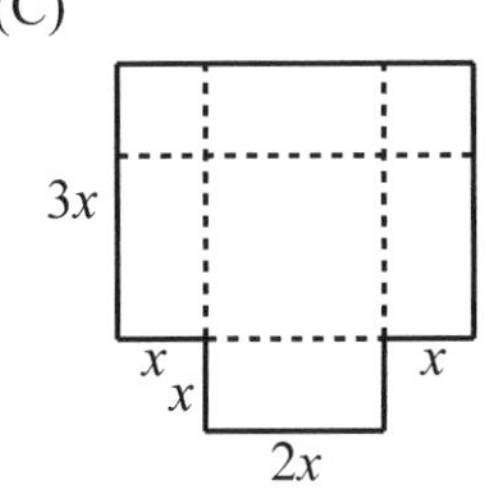

(D)

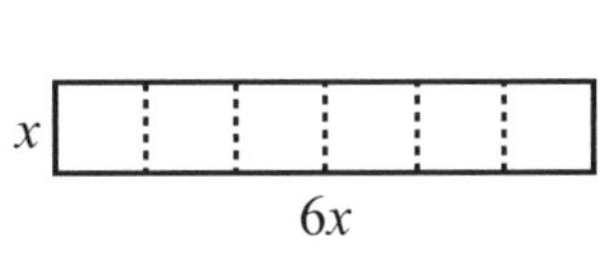

(E)

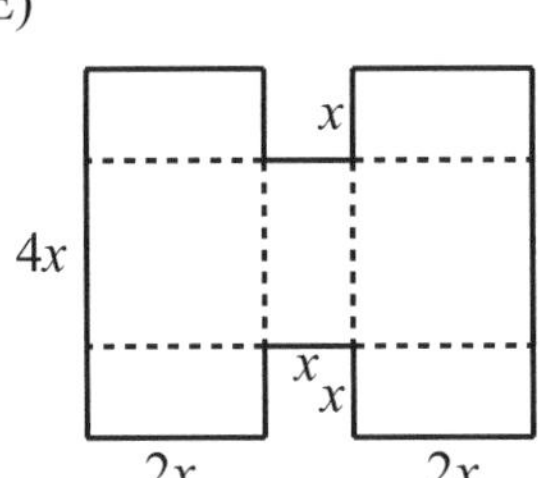

27. If $\log_a 4 = x$ and $a^{2x-6} = 4$, what is x?

(A) -6
(B) -2
(C) 2
(D) 4
(E) 6

GO ON TO THE NEXT PAGE

USE THIS SPACE FOR SCRATCHWORK.

28. If $ax^2 + bx + c = 0$ has the solution $-1 + (2\sqrt{6})\frac{i}{3}$, which of the following must also be a solution?

    (A) $-\frac{5}{2}$

    (B) $-\frac{6}{5}$

    (C) $-1 - (2\sqrt{6})\frac{i}{3}$

    (D) $1 + (2\sqrt{6})\frac{i}{3}$

    (E) $1 - (2\sqrt{6})\frac{i}{3}$

29. Joe and Jim find ten \$1 bills on the ground as they walk to school. In how many ways can they divide the \$10 between themselves if each of them is to get at least \$1?

    (A) 2
    (B) 5
    (C) 9
    (D) 10
    (E) 20

30. If $f(x, y, z) = (z, x, y)$ and $g(x, y, z) = (1, y, x)$, what is $g(f(1, y, z))$?

    (A) $(1, 1, x)$

    (B) $(1, 1, y)$

    (C) $(1, 1, z)$

    (D) $(1, y, 1)$

    (E) $(1, z, 1)$

GO ON TO THE NEXT PAGE

USE THIS SPACE FOR SCRATCHWORK.

31. If $x_n = \dfrac{1}{1+x_{n-1}}$ and $x_0 = 2$, what is the value of x_3?

(A) $\frac{7}{4}$
(B) $\frac{4}{3}$
(C) $\frac{3}{4}$
(D) $\frac{4}{7}$
(E) $\frac{7}{11}$

32. Suppose the graph of $f(x) = 2x^2$ is reflected about the x-axis, translated 2 units right, and then 5 units up. If the resulting graph represents $g(x)$, for what value of x does $g(x)$ reach its highest point?

(A) -3
(B) 0
(C) 1
(D) 2
(E) 5

33. In each of the following sets of numbers, the range exceeds the standard deviation, EXCEPT

(A) {1,2,3,4,5}
(B) {2,4,7,9,11}
(C) {2,5,6,6,10}
(D) {7,7,7,7,7}
(E) {11,13,14,14,15}

GO ON TO THE NEXT PAGE

USE THIS SPACE FOR SCRATCHWORK.

34. For all θ such that $\tan\theta$ is defined, $(\sin^2\theta - 1)(\tan^2\theta + 1) =$

(A) -1
(B) 0
(C) 1
(D) $\sin^2\theta\tan^2\theta - 1$
(E) $\sin^2\theta\tan^2\theta - \sec^2\theta - 1$

35. A cube is inscribed in a sphere of radius r. What is the ratio of the volume of the cube to the volume of the sphere?

(A) $1 : \pi$
(B) $2 : \pi$
(C) $2 : \pi\sqrt{3}$
(D) $2 : 3\pi$
(E) $1 : 4/3$

36. If $f(x) = 2 + \dfrac{x}{x-2}$ and $h(x) = \dfrac{2}{x-2} - x$, for what value(s) of x does $f(x) = h(x)$?

(A) ± 2
(B) 2
(C) ± 3
(D) 2 and -3
(E) -3

GO ON TO THE NEXT PAGE

USE THIS SPACE FOR SCRATCHWORK.

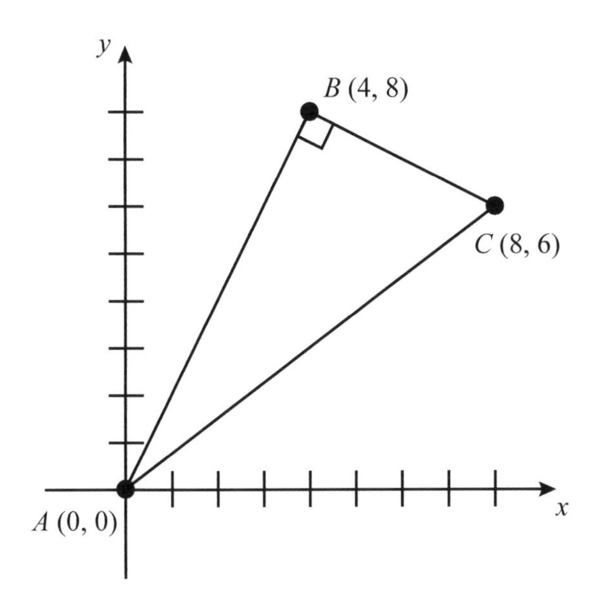

37. Triangle ABC is a right triangle. What could be the coordinates of point D (not shown) such that $\overline{AD} \cong \overline{BD}$?

(A) (3,4)
(B) (3.2,2.4)
(C) (3.6,2.7)
(D) (4,3)
(E) $(4, 3\sqrt{3})$

$$\frac{x^2}{9} - \frac{y^2}{16} = 1$$

38. Which of the following is an equation for an asymptote of the hyperbola represented by the above equation?

(A) $y = \frac{16}{9}x$
(B) $y = \frac{-9}{16}x$
(C) $y = \frac{3}{4}x$
(D) $y = \frac{4}{3}x$
(E) $y = x$

GO ON TO THE NEXT PAGE

USE THIS SPACE FOR SCRATCHWORK.

39. A rectangle has length l and width w. The perimeter of the rectangle is 96 inches. What is the greatest possible area of such a rectangle?

(A) 576 in^2
(B) 144 in^2
(C) 96 in^2
(D) 48 in^2
(E) 24 in^2

40. A string of beads is made up of a repeating sequence of one yellow, one red, one blue, and one green bead, in that order. What is the color of the 1025th bead?

(A) Yellow
(B) Red
(C) Blue
(D) Green
(E) It cannot be determined from the information given.

41. If $f(x) = 3\ln(x - 2)$ for $x > 2$, then $f^{-1}(x) =$

(A) $e^{\frac{x}{3}} + 2$
(B) $\frac{e^x}{3} + 2$
(C) $-\frac{\ln(x+2)}{3}$
(D) $2\ln\left(\frac{x}{3}\right)$
(E) $3e^{x-2}$

GO ON TO THE NEXT PAGE

USE THIS SPACE FOR SCRATCHWORK.

42. What is the range of the piecewise function defined below?

$$f(x) = \begin{cases} x^2 & \text{if } x > 2 \\ 3x - 3 & \text{if } x \leq 2 \end{cases}$$

(A) All real numbers
(B) $y > 4$
(C) $y \leq 3$
(D) $3 \leq y < 4$
(E) $y \leq 3$ or $y > 4$

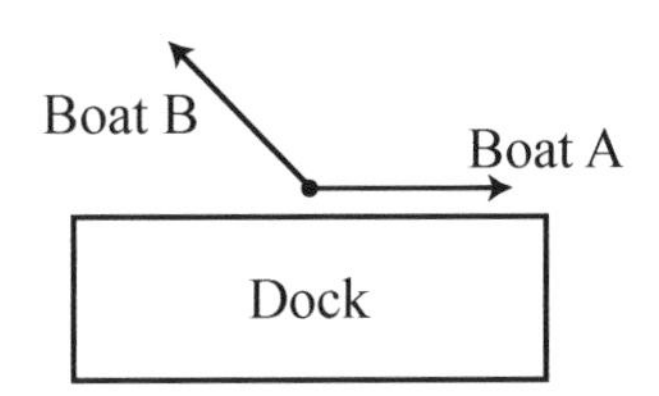

43. At 5pm a boat leaves the dock at 20mph traveling due east. Half an hour later a second boat leaves the dock travelling at 30mph, 45° northwest. At 8pm, what is the approximate distance between boats?

(A) 46 miles
(B) 84 miles
(C) 125 miles
(D) 133 miles
(E) 135 miles

44. What is the value of $\sum_{n=1}^{\infty} (2)^n$?

(A) -4
(B) -3
(C) 3
(D) 4
(E) ∞

GO ON TO THE NEXT PAGE

USE THIS SPACE FOR SCRATCHWORK.

45. If $(x - i)(x + i) = 17$, what is one possible value for x?

(A) 2
(B) 3
(C) $\sqrt{16 + 2i}$
(D) $\sqrt{16 - 2i}$
(E) 4

46. Consider the graph of $|x| - |y| = 1$. Which of the following restrictions would result in the graph of a function?

(A) $x \geq 0$
(B) $x \leq 0$
(C) $|x| \geq 1$
(D) $y \geq 0$
(E) $|y| \geq 1$

47. $\lim_{x \to 3} \left\{ \begin{array}{ll} 2x & : x \leq 3 \\ -x + 6 & : x > 3 \end{array} \right. =$

(A) 0
(B) 1
(C) 3
(D) 6
(E) The limit does not exist.

48. The position of a bug crawling on a coordinate plane is given by the parametric equations $x = 2t^2 + 7t$ and $y = t^3 - t$. At $t = 4$, how far is the bug from its starting point at t=0?

(A) 60
(B) $60\sqrt{2}$
(C) $60\sqrt{3}$
(D) 120
(E) 360

GO ON TO THE NEXT PAGE

USE THIS SPACE FOR SCRATCHWORK.

49. How many zeros within the domain $\frac{-7\pi}{8} < x < \frac{7\pi}{8}$ does $y = 2\tan(4x) - 3$ have?

(A) 1
(B) 2
(C) 4
(D) 7
(E) 8

50. Let $A = \begin{pmatrix} 0 & 1 \\ 1 & 0 \end{pmatrix}$ be a 2×2 matrix. If $v = (x, y)$ is a point in the xy plane, which of the following always describes the product vA?

(A) Rotation by 90°
(B) Reflection over the line $y = x$
(C) Reflection over the line $y = -x$
(D) Translation of 1 unit horizontally and 1 unit vertically
(E) Rotation by 45°

STOP

If you finish before time is up, you may check your work on the rest of the test.

SCORING YOUR TEST

A word of caution

- This presumably is the third test you've taken, so good job,[1] but maybe also go get an ice cream sundae or something.[2]

The rest of it

- Compare the answers you entered on your answer sheet with the answer key chart on the facing page and mark each answer correct, incorrect, or skipped.
- Use the formula at the bottom of the page to calculate your raw score.
- Turn the page to get your scaled score, which (not to belabor the point) may differ from the score you receive on your official test.

---

1 Our records indicate that you're quite studious.

2 Breaks are good.

YOUR ANSWERS

QUESTION NUMBER	CORRECT ANSWER	CORRECT	INCORRECT	SKIP
1	A			
2	A			
3	C			
4	A			
5	D			
6	D			
7	B			
8	A			
9	B			
10	B			
11	B			
12	B			
13	A			
14	C			
15	B			
16	E			
17	D			
18	B			
19	D			
20	D			
21	D			
22	B			
23	C			
24	C			
25	D			

YOUR ANSWERS

QUESTION NUMBER	CORRECT ANSWER	CORRECT	INCORRECT	SKIP
26	A			
27	E			
28	C			
29	C			
30	C			
31	D			
32	D			
33	D			
34	A			
35	C			
36	E			
37	D			
38	D			
39	A			
40	A			
41	A			
42	E			
43	C			
44	E			
45	E			
46	D			
47	E			
48	B			
49	D			
50	B			

TOTAL CORRECT		TOTAL INCORRECT		RAW SCORE
	- .25 ×		=	

Finding your scaled score

- Round your raw score to the nearest digit.
- Look up the corresponding scaled score in the table below.
- Cast off the shackles of fate and revel in your radical autonomy.

RAW SCORE	SCALED SCORE	RAW SCORE	SCALED SCORE	RAW SCORE	SCALED SCORE
50	800	29	660	8	490
49	800	28	650	7	480
48	800	27	640	6	480
47	800	26	630	5	470
46	800	25	630	4	460
45	800	24	620	3	450
44	800	23	610	2	440
43	800	22	600	1	430
42	790	21	590	0	410
41	780	20	580	-1	390
40	770	19	570	-2	370
39	760	18	560	-3	360
38	750	17	560	-4	340
37	740	16	550	-5	340
36	730	15	540	-6	330
35	720	14	530	-7	320
34	710	13	530	-8	320
33	700	12	520	-9	320
32	690	11	510	-10	320
31	680	10	500	-11	310
30	670	9	500	-12	310

TEST FOUR

AN *i* FOR AN *i*

For most accurate results, take this test in a setting similar to the one you'll take the real test in. We've organized the essentials into this simple mnemonic:

Leave noisy and crowded areas; find a secluded, tranquil test-taking spot.

If you haven't done so already, invest in a timer and set it to one hour.

Be your best by using a scientific or graphing calculator.

Every question you solve should be recorded on your answer sheet in the back of this book or downloaded from barebonesguide.com/answersheet.

Legally speaking, you can use outside resources on a practice test, but it's still not a good idea.

After you have finished, consult our scoring instructions at the end of the test to calculate your raw and scaled scores.

MATHEMATICS LEVEL 2 TEST

REFERENCE INFORMATION

YOU MAY USE THE FOLLOWING INFORMATION AS YOU TAKE THIS TEST.

A right circular cone with radius r and height h has volume $V = \frac{1}{3}\pi r^2 h$

A right circular cone with circumference of the base c and slant height ℓ has lateral area $S = \frac{1}{2}c\ell$

A sphere with radius r has volume $V = \frac{4}{3}\pi r^3$

A sphere with radius r has surface area $S = 4\pi r^2$

A pyramid with base area B and height h has volume $V = \frac{1}{3}Bh$

GO ON TO THE NEXT PAGE

MATHEMATICS LEVEL 2 TEST

Choose the best choice given for each of the following questions. The exact value may not be one of the choices; in this case select the best approximation of the true value.

Note the following items carefully.

(1) You will need a scientific or graphing calculator to answer some of the questions in this test. Some of these questions will require you to determine whether your calculator should be in degree or radian mode. For other questions, however, you will not need a calculator at all.

(2) All figures are drawn to scale and lie in the plane unless otherwise specified.

(3) For the purposes of this test, the domain of any function f is the set of real numbers x for which $f(x)$ is a real number and the range of f is the set of real numbers $f(x)$, where x is in the domain of f. Any deviation from these assumptions will be noted in each problem to which it applies.

USE THIS SPACE FOR SCRATCHWORK.

1. What is the midpoint between $(-1, 2)$ and $(5, 6)$?

(A) $(-6, -4)$
(B) $(-3, -2)$
(C) $(-2, 4)$
(D) $(2, 4)$
(E) $(3, 3)$

2. The number of miles m driven in a car is given by $m = 12g + 6$, where g is the number of gallons of gas in a tank. The amount spent on maintenance, n, is determined by how many miles the car has driven in the equation $n = \frac{10m}{3} - 40$, where $n > 20$. Which equation expresses the relation between number of gallons g and maintenance costs n?

(A) $n = 40g - 38$
(B) $n = 40g - 20$
(C) $n = g - 1$
(D) $n = 120g - 34$
(E) $n = 4g - 34$

GO ON TO THE NEXT PAGE

USE THIS SPACE FOR SCRATCHWORK.

3. If $a + b = 8$ and $b - c = 7$ and $a - b + c = -3$, what is the value of b?

(A) 3
(B) 4
(C) 5
(D) 6
(E) 7

4. If $f(x) = x^2 + 2$ and $g(x) = \sqrt{x - 1}$, then $f(g(10)) =$

(A) $\sqrt{101}$
(B) $\sqrt{103}$
(C) 10
(D) 11
(E) 12

5. If plane A is parallel to plane B, plane C is parallel to plane D, and plane B is perpendicular to plane D, which of the following must be true?

I. Plane D is perpendicular to Plane A.
II. Plane A is perpendicular to Plane C.
III. Plane A is parallel to Plane C.

(A) None of the above
(B) I only
(C) I and II only
(D) I and III only
(E) I, II, and III

GO ON TO THE NEXT PAGE

USE THIS SPACE FOR SCRATCHWORK.

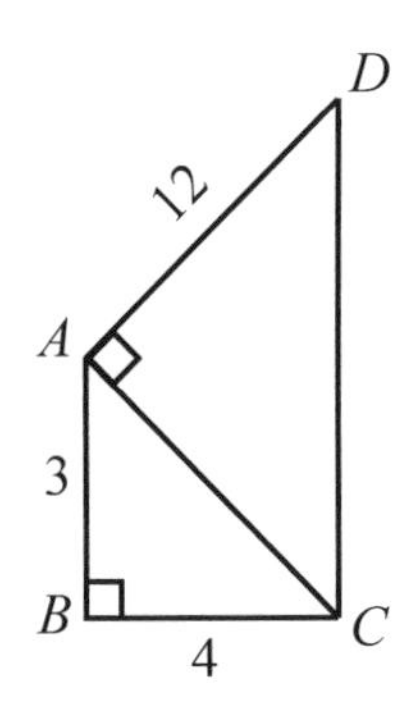

6. What is the measure of $\angle ACD$?

(A) 23.6°
(B) 33.4°
(C) 44.3°
(D) 46.2°
(E) 67.4°

7. What is the value of x^{-1} if $x = \dfrac{1}{10^2 - 8^2}$?

(A) $\dfrac{1}{6}$
(B) $\dfrac{1}{36}$
(C) 2
(D) 6
(E) 36

8. The distance between point $A(x, y)$ and point $B(-x, y)$ is z, and the midpoint between A and B is located at point C. What is the distance between C and the origin?

(A) 0
(B) $\dfrac{z}{2}$
(C) y
(D) $\dfrac{y}{z}$
(E) $\dfrac{y}{2}$

GO ON TO THE NEXT PAGE

USE THIS SPACE FOR SCRATCHWORK.

9. If $h(x) = (x - 1)(x + 1)$ and $f(x) = \sqrt{x^2 - 1}$ and $h(x) = f(x) \cdot g(x)$, then $g(x) =$

 (A) $x + 1$
 (B) $x - 1$
 (C) $x^2 - 1$
 (D) $\sqrt{x^2 - 1}$
 (E) $\sqrt{x - 1}$

10. If $4x - 12 = \frac{K}{2}(3 - x)$, then $K =$

 (A) $\frac{1}{12}$
 (B) 8
 (C) 4
 (D) -8
 (E) -12

11. If $(1 - \sin^2\theta) = \frac{1}{2}$, then $\cos\theta =$

 (A) $\frac{1}{2}$
 (B) $\frac{1}{4}$
 (C) $45°$
 (D) $60°$
 (E) $\frac{\sqrt{2}}{2}$

GO ON TO THE NEXT PAGE

USE THIS SPACE FOR SCRATCHWORK.

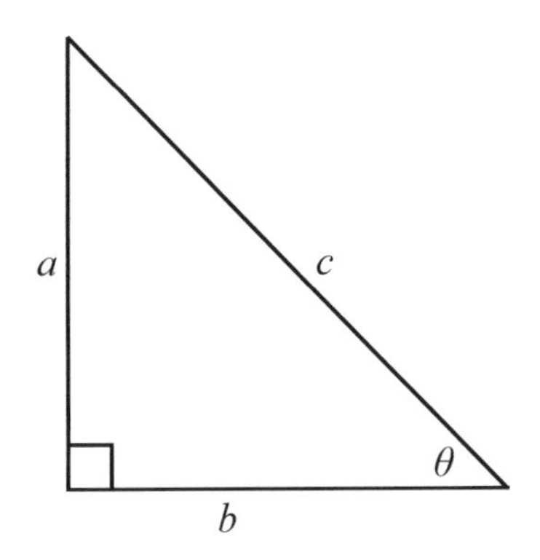

12. According to the figure, which of the following gives the measure of θ?

    (A) $\sin^{-1}\left(\frac{a}{b}\right)$
    (B) $\cos^{-1}\left(\frac{a}{b}\right)$
    (C) $\cos^{-1}\left(\frac{a}{c}\right)$
    (D) $\sin^{-1}\left(\frac{a}{c}\right)$
    (E) $\tan^{-1}\left(\frac{b}{a}\right)$

13. What is the equation of the vertical asymptote of $f(x) = \log(2x)$?

    (A) $x = 2$
    (B) $x = 1$
    (C) $x = 0$
    (D) $x = 4$
    (E) $x = -2$

GO ON TO THE NEXT PAGE

USE THIS SPACE FOR SCRATCHWORK.

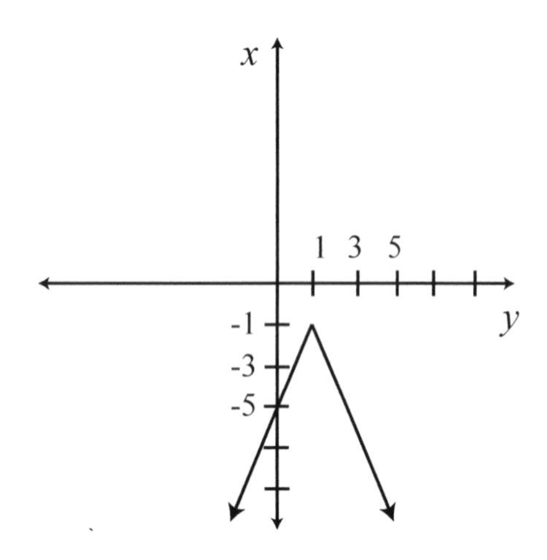

14. In the graph of $y = -2\,|\,x - 1\,| + c$ shown above, what is the value of c?

 (A) 4
 (B) 3
 (C) 0
 (D) -1
 (E) -2

15. If $\ln x = 3$, then $e^x =$

 (A) 3
 (B) e
 (C) e^3
 (D) 271
 (E) 528, 491, 311

16. The average age of 5 people is 22 years old. Two people join the group and the average becomes 24. What is the average age of the two who joined the group?

 (A) 24
 (B) 25
 (C) 26
 (D) 28
 (E) 29

GO ON TO THE NEXT PAGE

USE THIS SPACE FOR SCRATCHWORK.

17. Line p has a slope of 0 and passes through the point (3,2). Which of the following equations describes a line parallel to line p?

(A) $y = 3$
(B) $x = 3$
(C) $x = 2$
(D) $y = 2x$
(E) $y = 3x$

18. If $a < x - 3$ and $2x - 6 < b$, which of the following must be true?

(A) $a < x - 3 < \frac{b}{2}$
(B) $a < b$
(C) $a < x - 3 < 2b$
(D) $2a < x - 3 < 2b$
(E) $a + 6 < x < b$

19. A line p is drawn tangent to the function $f(x) = 2^x$ at some point $(x, f(x))$. Which of the following must be true?

(A) The slope of line p equals zero.
(B) The slope of line p is positive.
(C) The slope of line p is negative.
(D) The y intercept of line p is positive.
(E) The x intercept of line p is negative.

GO ON TO THE NEXT PAGE

USE THIS SPACE FOR SCRATCHWORK.

20. If $f(a) = f(b) = c$, where a, b, and c are distinct real numbers, which of the following must be true?

I. $f(a + b) = 2c$
II. The reflection of f across the line $y = x$ is not a function.
III. $f(f(a)) = b$

(A) I only
(B) II only
(C) I and II only
(D) II and III only
(E) I, II, and III

21. In a survey of 50 people, 26 responded that they use brand A, 28 responded that they use brand B, and 6 people said they don't used either brand. What is the probability that a person chosen at random from these respondents uses both brands?

(A) 0.12
(B) 0.20
(C) 0.25
(D) 0.52
(E) 0.56

22. A right triangle has legs of length 5 and $5\sqrt{2}$. To the nearest tenth of a degree, what is the measure of the smallest angle in this triangle?

(A) 35.3°
(B) 41.8°
(C) 54.7°
(D) 48.2°
(E) 90°

GO ON TO THE NEXT PAGE

USE THIS SPACE FOR SCRATCHWORK.

23. If a, b, c, and d are real integers, and if $f(x) = ax^3 + bx^2 + cx - d$, which of the following must be true?

    (A) The function f has at most 2 non-real roots.
    (B) $(d, f(d))$ represents the function's minimum value.
    (C) All roots of f are positive.
    (D) The function f has no real roots.
    (E) The function f has 2 rational roots.

24. What is the domain of $f(x) = \frac{\sqrt{x+2}}{\sqrt{x-1}}$?

    (A) $x \geq -2$
    (B) $x > -2$
    (C) $x \geq 1$
    (D) $x > 1$
    (E) $x < -2$ or $x \geq 1$

25. Which of the following polar equations can be written in rectangular form as $x^2 + y^2 = 9$?

    (A) $r = 3$
    (B) $r = \sin\theta$
    (C) $r = 2\sin\theta$
    (D) $r = \sin 3\theta$
    (E) $r = 3\theta$

26. If $g(x) = (3 - x)^2$ and $f(x) = \sqrt{7 + 4ix}$, what does $g\left(f\left(\frac{-i}{2}\right)\right)$ equal?

    (A) $14 - 6\sqrt{5}$
    (B) $\frac{i}{2}$
    (C) $7 - 2i$
    (D) $\sqrt{5}$
    (E) 0

GO ON TO THE NEXT PAGE

27. In January 1988 a tree had a height of 40 feet. If it grew at a rate of 3% each year, in what year did it reach a height of 50 feet?

(A) 1991
(B) 1995
(C) 1999
(D) 2000
(E) 2007

28. For which of the following values of θ is it always true that $\cos(\theta) = \sin(90° - \theta)$?

(A) $0° < \theta < 90°$ only
(B) $0° < \theta < 180°$ only
(C) $90° < \theta < 270°$ only
(D) $0° < \theta < 90°$ or $180° < \theta < 360°$
(E) All real values of θ

GO ON TO THE NEXT PAGE

USE THIS SPACE FOR SCRATCHWORK.

29. In how many ways can five people stand in a line so that the tallest person is not standing at either end?

    (A) 24
    (B) 36
    (C) 48
    (D) 52
    (E) 72

When it rains, Janet goes to the movies.

30. Which of the following must be true if the statement above is true?

    (A) When Janet goes to the movies, it rains.
    (B) When it doesn't rain, Janet doesn't go to the movies.
    (C) When it doesn't rain, Janet goes to the movies.
    (D) When Janet doesn't go to the movies, it doesn't rain.
    (E) When it rains, Janet doesn't go to the movies.

31. If $f(x) = \sqrt[3]{x+2}$, for how many values of x does $f(x) = f(-x)$?

    (A) None
    (B) One
    (C) Two
    (D) Three
    (E) An infinite number

32. What value does $\frac{3}{x^2 - 4}$ approach as x approaches 2?

    (A) 0
    (B) 1
    (C) 2
    (D) 3
    (E) The limit does not exist.

GO ON TO THE NEXT PAGE

USE THIS SPACE FOR SCRATCHWORK.

33. A sector of $40°$ is cut competely out of a circle of radius 4. Approximately what is the perimeter of the part of the circle that remains?

(A) 30.34
(B) 22.34
(C) 18.34
(D) 16.34
(E) 2.8

$x =$	0	1	2	3
$f(x)$	-1	0	1	2
$g(x)$	2	1	0	-1

34. In the table above, function values are displayed for different values of x. If f and g have domain $\{0, 1, 2, 3\}$, $g(a) = h$, and $f(h) = 1$, what is the value of $g(a + h)$?

(A) $f(3)$
(B) $f(-1)$
(C) $g(1)$
(D) $f(1)$
(E) $f(0)$

35. What is the lateral surface area of a cylinder with radius 3 and volume 36π?

(A) 12π
(B) 16π
(C) 24π
(D) 36π
(E) 42π

36. For $f(x) = \tan(2x)$, which describes the complete set of x-values for which f is undefined?

(A) $\frac{\pi}{2} \pm k\pi$, where k is an integer
(B) $\frac{\pi}{4} \pm \frac{\pi}{2}k$, where k is an integer
(C) $\pi \pm \frac{\pi}{2}k$, where k is an integer
(D) $\frac{\pi}{2} \pm \pi k$, where k is an integer
(E) $\frac{\pi}{4} \pm \pi k$ where k is an integer

GO ON TO THE NEXT PAGE

USE THIS SPACE FOR SCRATCHWORK.

37. If $g(x) = \frac{x}{x+1}$ what is $g^{-1}(2)$?

(A) -2
(B) -1
(C) 1
(D) 2
(E) 4

38. Within the domain $\frac{\pi}{2} < \theta < \pi$, if $\sin\theta = a\cos\theta$, a may be equal to

(A) -1
(B) 0
(C) 1
(D) 2
(E) 4

39. For $f(x) = \frac{-3x^3 + 2x - 1}{4x^3 - x}$, $f(3^{100}) - f(-3^{99}) =$

(A) 0
(B) -1.5
(C) 1.5
(D) 3
(E) 6

64, 32, 48, 40, 44, 42,….

40. For the sequence above, which could be the recursive definition to find any term, A_n?

(A) $A_0 = 64, A_1 = 32, A_n = A_{n-1}(0.5)$
(B) $A_0 = 64, A_1 = 32, A_n = A_{n-2}\left(\frac{2}{3}\right)$
(C) $A_0 = 64, A_1 = 32, A_n = \frac{A_{n-1} + A_{n-2}}{2}$
(D) $A_0 = 128, A_1 = 96, A_n = \frac{A_{n-1} + A_{n-2}}{2}$
(E) $A_0 = 96, A_1 = 64, A_n = A_{n-1}\left(\frac{2}{3}\right)$

GO ON TO THE NEXT PAGE

USE THIS SPACE FOR SCRATCHWORK.

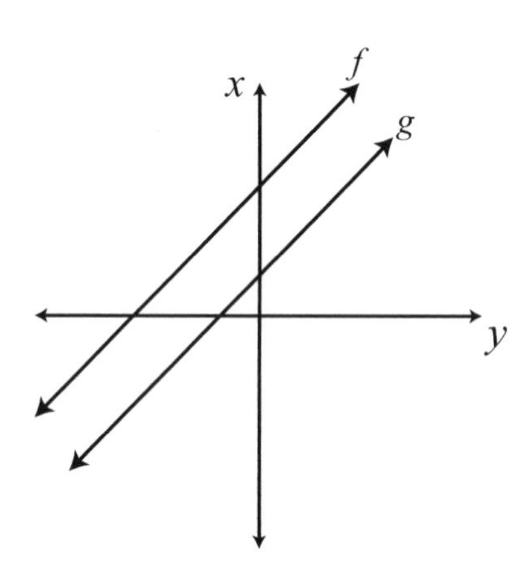

41. *f* and *g* are functions drawn above. If *f* is parallel to *g*, which could be the graph of $\frac{f}{g}(x)$?

(A) 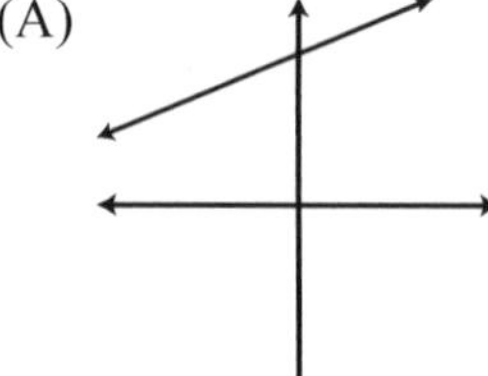

(B) 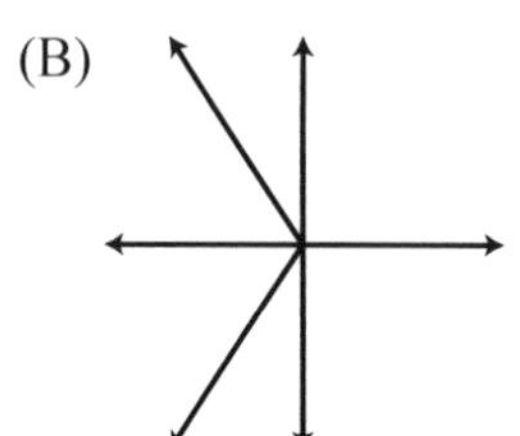

(C) 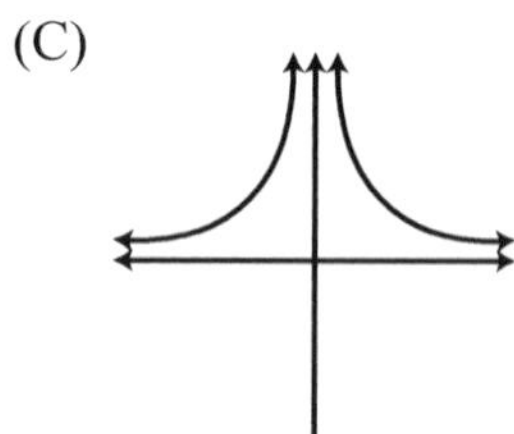

(D) 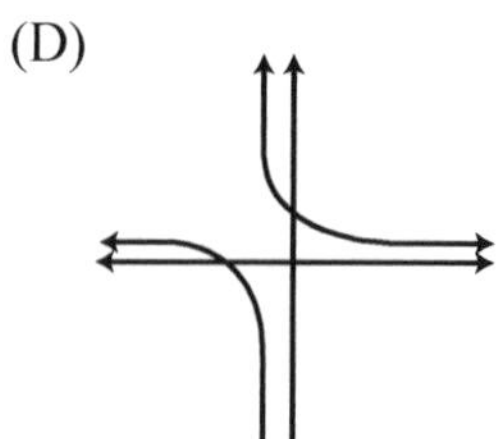

(E) 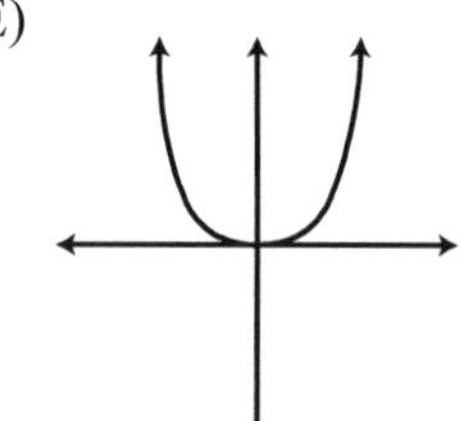

GO ON TO THE NEXT PAGE

USE THIS SPACE FOR SCRATCHWORK.

42. Set A contains all points (x, y) that are a distance of $\sqrt{2}$ from the point (3, 5). One of the elements (a, b) in the set is chosen at random. Which of the following must be true of (a, b)?

I. Both a and b are integers.
II. Both a and b are rational numbers.
III. Both a and b are irrational numbers.

(A) None of the above
(B) I only
(C) II only
(D) III only
(E) I and II only

43. A man on a bike rode for 30 minutes at 20 mph from point A due east to point B. Then he rode for 40 minutes at 30 mph due north from point B to point C. Then he rode for 20 minutes at 15 mph due east again from point C to point D. How many fewer minutes would the trip have taken if he had traveled in a straight line from A to D at 25 mph?

(A) 0 minutes
(B) 20 minutes
(C) 30 minutes
(D) 35.7 minutes
(E) 45 minutes

44. If g is a third degree polynomial function with roots 12 and -6, and $g(5) < 0 < g(9)$, which of the following could be a root of g?

(A) 0
(B) 5
(C) 7
(D) 9
(E) 14

GO ON TO THE NEXT PAGE

USE THIS SPACE FOR SCRATCHWORK.

45. $\sum_{k=1}^{4} 3k^{(k-1)} =$

(A) 226
(B) 228
(C) 331
(D) 602
(E) 804

46. On any given day, there is a 40% chance of rain in the Northeast. If the probability of rain on any given day is independent of the probability of rain on any other day, what is the probability that it rains at least once in the Northeast during a four-day period?

(A) 8%
(B) 10%
(C) 50%
(D) 78%
(E) 87%

47. At what point do the ellipses described by the equations $\frac{x^2}{4} + \frac{y^2}{1} = 1$ and $\frac{(x-4)^2}{4} + \frac{y^2}{1} = 1$ intersect?

(A) $(0, 1)$
(B) $(1, 0)$
(C) $(2, 0)$
(D) $(2, 1)$
(E) $(2, 2)$

48. The third term of a geometric series is 9, and the fifth term is 1. If the common ratio between terms is a positive number, approximately what is the sum of the first 10,000 terms?

(A) 121.5
(B) 100
(C) 81.5
(D) 27
(E) 13.5

GO ON TO THE NEXT PAGE

USE THIS SPACE FOR SCRATCHWORK.

49. Vector A with magnitude 6 makes a 40° angle with vector B. The resultant vector C, the sum of vectors A and B, has a magnitude of 12.22. What is the magnitude of vector B?

    (A) 5
    (B) 6
    (C) 6.5
    (D) 7
    (E) 8

50. Two similar prisms have lateral areas of 48 and 75 respectively. What is the ratio of their volumes?

    (A) 4 : 5
    (B) $4\sqrt{3}$: 5
    (C) 16 : 25
    (D) 64 : 125
    (E) It cannot be determined from the information given.

STOP

If you finish before time is up, you may check your work on the rest of the test.

SCORING YOUR TEST

Congratulations

- We're all quite impressed[1] that you've gotten this far.

Etc.

- Do the thing to get your raw score.[2]
- Understand that we would guarantee your official score if we could, but we can't, so we don't.[3]
- Turn the page.

1 And a little concerned. But mostly impressed.

2 If you skipped tests to take this one, maybe look back at previous tests to get the directions we wrote before we got bored.

3 That's life for you.

QUESTION NUMBER	CORRECT ANSWER	YOUR ANSWERS CORRECT	YOUR ANSWERS INCORRECT	YOUR ANSWERS SKIP
1	D			
2	B			
3	B			
4	D			
5	C			
6	E			
7	E			
8	C			
9	D			
10	D			
11	E			
12	D			
13	C			
14	D			
15	E			
16	E			
17	A			
18	A			
19	B			
20	B			
21	B			
22	A			
23	A			
24	D			
25	A			

QUESTION NUMBER	CORRECT ANSWER	YOUR ANSWERS CORRECT	YOUR ANSWERS INCORRECT	YOUR ANSWERS SKIP
26	E			
27	B			
28	E			
29	E			
30	D			
31	B			
32	E			
33	A			
34	D			
35	C			
36	B			
37	A			
38	A			
39	A			
40	C			
41	D			
42	A			
43	C			
44	C			
45	B			
46	E			
47	C			
48	A			
49	D			
50	D			

TOTAL CORRECT [ ] - .25 × **TOTAL INCORRECT** [ ] = **RAW SCORE** [ ]

Finding your scaled score

- It's right there in the table.[1]

RAW SCORE	SCALED SCORE
50	800
49	800
48	800
47	800
46	800
45	800
44	800
43	800
42	790
41	780
40	770
39	760
38	750
37	740
36	730
35	720
34	710
33	700
32	690
31	680
30	670

RAW SCORE	SCALED SCORE
29	660
28	650
27	640
26	630
25	630
24	620
23	610
22	600
21	590
20	580
19	570
18	560
17	560
16	550
15	540
14	530
13	530
12	520
11	510
10	500
9	500

RAW SCORE	SCALED SCORE
8	490
7	480
6	480
5	470
4	460
3	450
2	440
1	430
0	410
-1	390
-2	370
-3	360
-4	340
-5	340
-6	330
-7	320
-8	320
-9	320
-10	320
-11	310
-12	310

1 We have confidence in your ability to find it.

ANSWER SHEETS

TEST:

DATE:

1 (A) (B) (C) (D) (E)
2 (A) (B) (C) (D) (E)
3 (A) (B) (C) (D) (E)
4 (A) (B) (C) (D) (E)
5 (A) (B) (C) (D) (E)
6 (A) (B) (C) (D) (E)
7 (A) (B) (C) (D) (E)
8 (A) (B) (C) (D) (E)
9 (A) (B) (C) (D) (E)
10 (A) (B) (C) (D) (E)
11 (A) (B) (C) (D) (E)
12 (A) (B) (C) (D) (E)
13 (A) (B) (C) (D) (E)
14 (A) (B) (C) (D) (E)
15 (A) (B) (C) (D) (E)
16 (A) (B) (C) (D) (E)
17 (A) (B) (C) (D) (E)
18 (A) (B) (C) (D) (E)
19 (A) (B) (C) (D) (E)
20 (A) (B) (C) (D) (E)
21 (A) (B) (C) (D) (E)
22 (A) (B) (C) (D) (E)
23 (A) (B) (C) (D) (E)
24 (A) (B) (C) (D) (E)
25 (A) (B) (C) (D) (E)
26 (A) (B) (C) (D) (E)
27 (A) (B) (C) (D) (E)
28 (A) (B) (C) (D) (E)
29 (A) (B) (C) (D) (E)
30 (A) (B) (C) (D) (E)
31 (A) (B) (C) (D) (E)
32 (A) (B) (C) (D) (E)
33 (A) (B) (C) (D) (E)
34 (A) (B) (C) (D) (E)
35 (A) (B) (C) (D) (E)
36 (A) (B) (C) (D) (E)
37 (A) (B) (C) (D) (E)
38 (A) (B) (C) (D) (E)
39 (A) (B) (C) (D) (E)
40 (A) (B) (C) (D) (E)
41 (A) (B) (C) (D) (E)
42 (A) (B) (C) (D) (E)
43 (A) (B) (C) (D) (E)
44 (A) (B) (C) (D) (E)
45 (A) (B) (C) (D) (E)
46 (A) (B) (C) (D) (E)
47 (A) (B) (C) (D) (E)
48 (A) (B) (C) (D) (E)
49 (A) (B) (C) (D) (E)
50 (A) (B) (C) (D) (E)
51 (A) (B) (C) (D) (E)
52 (A) (B) (C) (D) (E)
53 (A) (B) (C) (D) (E)
54 (A) (B) (C) (D) (E)
55 (A) (B) (C) (D) (E)
56 (A) (B) (C) (D) (E)
57 (A) (B) (C) (D) (E)
58 (A) (B) (C) (D) (E)
59 (A) (B) (C) (D) (E)
60 (A) (B) (C) (D) (E)
61 (A) (B) (C) (D) (E)
62 (A) (B) (C) (D) (E)
63 (A) (B) (C) (D) (E)
64 (A) (B) (C) (D) (E)
65 (A) (B) (C) (D) (E)
66 (A) (B) (C) (D) (E)
67 (A) (B) (C) (D) (E)
68 (A) (B) (C) (D) (E)
69 (A) (B) (C) (D) (E)
70 (A) (B) (C) (D) (E)
71 (A) (B) (C) (D) (E)
72 (A) (B) (C) (D) (E)
73 (A) (B) (C) (D) (E)
74 (A) (B) (C) (D) (E)
75 (A) (B) (C) (D) (E)
76 (A) (B) (C) (D) (E)
77 (A) (B) (C) (D) (E)
78 (A) (B) (C) (D) (E)
79 (A) (B) (C) (D) (E)
80 (A) (B) (C) (D) (E)
81 (A) (B) (C) (D) (E)
82 (A) (B) (C) (D) (E)
83 (A) (B) (C) (D) (E)
84 (A) (B) (C) (D) (E)
85 (A) (B) (C) (D) (E)
86 (A) (B) (C) (D) (E)
87 (A) (B) (C) (D) (E)
88 (A) (B) (C) (D) (E)
89 (A) (B) (C) (D) (E)
90 (A) (B) (C) (D) (E)
91 (A) (B) (C) (D) (E)
92 (A) (B) (C) (D) (E)
93 (A) (B) (C) (D) (E)
94 (A) (B) (C) (D) (E)
95 (A) (B) (C) (D) (E)
96 (A) (B) (C) (D) (E)
97 (A) (B) (C) (D) (E)
98 (A) (B) (C) (D) (E)
99 (A) (B) (C) (D) (E)
100 (A) (B) (C) (D) (E)

TEST:

DATE:

1 A B C D E	21 A B C D E	41 A B C D E	61 A B C D E	81 A B C D E
2 A B C D E	22 A B C D E	42 A B C D E	62 A B C D E	82 A B C D E
3 A B C D E	23 A B C D E	43 A B C D E	63 A B C D E	83 A B C D E
4 A B C D E	24 A B C D E	44 A B C D E	64 A B C D E	84 A B C D E
5 A B C D E	25 A B C D E	45 A B C D E	65 A B C D E	85 A B C D E
6 A B C D E	26 A B C D E	46 A B C D E	66 A B C D E	86 A B C D E
7 A B C D E	27 A B C D E	47 A B C D E	67 A B C D E	87 A B C D E
8 A B C D E	28 A B C D E	48 A B C D E	68 A B C D E	88 A B C D E
9 A B C D E	29 A B C D E	49 A B C D E	69 A B C D E	89 A B C D E
10 A B C D E	30 A B C D E	50 A B C D E	70 A B C D E	90 A B C D E
11 A B C D E	31 A B C D E	51 A B C D E	71 A B C D E	91 A B C D E
12 A B C D E	32 A B C D E	52 A B C D E	72 A B C D E	92 A B C D E
13 A B C D E	33 A B C D E	53 A B C D E	73 A B C D E	93 A B C D E
14 A B C D E	34 A B C D E	54 A B C D E	74 A B C D E	94 A B C D E
15 A B C D E	35 A B C D E	55 A B C D E	75 A B C D E	95 A B C D E
16 A B C D E	36 A B C D E	56 A B C D E	76 A B C D E	96 A B C D E
17 A B C D E	37 A B C D E	57 A B C D E	77 A B C D E	97 A B C D E
18 A B C D E	38 A B C D E	58 A B C D E	78 A B C D E	98 A B C D E
19 A B C D E	39 A B C D E	59 A B C D E	79 A B C D E	99 A B C D E
20 A B C D E	40 A B C D E	60 A B C D E	80 A B C D E	100 A B C D E

TEST:

DATE:

1 A B C D E	21 A B C D E	41 A B C D E	61 A B C D E	81 A B C D E
2 A B C D E	22 A B C D E	42 A B C D E	62 A B C D E	82 A B C D E
3 A B C D E	23 A B C D E	43 A B C D E	63 A B C D E	83 A B C D E
4 A B C D E	24 A B C D E	44 A B C D E	64 A B C D E	84 A B C D E
5 A B C D E	25 A B C D E	45 A B C D E	65 A B C D E	85 A B C D E
6 A B C D E	26 A B C D E	46 A B C D E	66 A B C D E	86 A B C D E
7 A B C D E	27 A B C D E	47 A B C D E	67 A B C D E	87 A B C D E
8 A B C D E	28 A B C D E	48 A B C D E	68 A B C D E	88 A B C D E
9 A B C D E	29 A B C D E	49 A B C D E	69 A B C D E	89 A B C D E
10 A B C D E	30 A B C D E	50 A B C D E	70 A B C D E	90 A B C D E
11 A B C D E	31 A B C D E	51 A B C D E	71 A B C D E	91 A B C D E
12 A B C D E	32 A B C D E	52 A B C D E	72 A B C D E	92 A B C D E
13 A B C D E	33 A B C D E	53 A B C D E	73 A B C D E	93 A B C D E
14 A B C D E	34 A B C D E	54 A B C D E	74 A B C D E	94 A B C D E
15 A B C D E	35 A B C D E	55 A B C D E	75 A B C D E	95 A B C D E
16 A B C D E	36 A B C D E	56 A B C D E	76 A B C D E	96 A B C D E
17 A B C D E	37 A B C D E	57 A B C D E	77 A B C D E	97 A B C D E
18 A B C D E	38 A B C D E	58 A B C D E	78 A B C D E	98 A B C D E
19 A B C D E	39 A B C D E	59 A B C D E	79 A B C D E	99 A B C D E
20 A B C D E	40 A B C D E	60 A B C D E	80 A B C D E	100 A B C D E

TEST:

DATE:

1 (A) (B) (C) (D) (E)	21 (A) (B) (C) (D) (E)	41 (A) (B) (C) (D) (E)	61 (A) (B) (C) (D) (E)	81 (A) (B) (C) (D) (E)
2 (A) (B) (C) (D) (E)	22 (A) (B) (C) (D) (E)	42 (A) (B) (C) (D) (E)	62 (A) (B) (C) (D) (E)	82 (A) (B) (C) (D) (E)
3 (A) (B) (C) (D) (E)	23 (A) (B) (C) (D) (E)	43 (A) (B) (C) (D) (E)	63 (A) (B) (C) (D) (E)	83 (A) (B) (C) (D) (E)
4 (A) (B) (C) (D) (E)	24 (A) (B) (C) (D) (E)	44 (A) (B) (C) (D) (E)	64 (A) (B) (C) (D) (E)	84 (A) (B) (C) (D) (E)
5 (A) (B) (C) (D) (E)	25 (A) (B) (C) (D) (E)	45 (A) (B) (C) (D) (E)	65 (A) (B) (C) (D) (E)	85 (A) (B) (C) (D) (E)
6 (A) (B) (C) (D) (E)	26 (A) (B) (C) (D) (E)	46 (A) (B) (C) (D) (E)	66 (A) (B) (C) (D) (E)	86 (A) (B) (C) (D) (E)
7 (A) (B) (C) (D) (E)	27 (A) (B) (C) (D) (E)	47 (A) (B) (C) (D) (E)	67 (A) (B) (C) (D) (E)	87 (A) (B) (C) (D) (E)
8 (A) (B) (C) (D) (E)	28 (A) (B) (C) (D) (E)	48 (A) (B) (C) (D) (E)	68 (A) (B) (C) (D) (E)	88 (A) (B) (C) (D) (E)
9 (A) (B) (C) (D) (E)	29 (A) (B) (C) (D) (E)	49 (A) (B) (C) (D) (E)	69 (A) (B) (C) (D) (E)	89 (A) (B) (C) (D) (E)
10 (A) (B) (C) (D) (E)	30 (A) (B) (C) (D) (E)	50 (A) (B) (C) (D) (E)	70 (A) (B) (C) (D) (E)	90 (A) (B) (C) (D) (E)
11 (A) (B) (C) (D) (E)	31 (A) (B) (C) (D) (E)	51 (A) (B) (C) (D) (E)	71 (A) (B) (C) (D) (E)	91 (A) (B) (C) (D) (E)
12 (A) (B) (C) (D) (E)	32 (A) (B) (C) (D) (E)	52 (A) (B) (C) (D) (E)	72 (A) (B) (C) (D) (E)	92 (A) (B) (C) (D) (E)
13 (A) (B) (C) (D) (E)	33 (A) (B) (C) (D) (E)	53 (A) (B) (C) (D) (E)	73 (A) (B) (C) (D) (E)	93 (A) (B) (C) (D) (E)
14 (A) (B) (C) (D) (E)	34 (A) (B) (C) (D) (E)	54 (A) (B) (C) (D) (E)	74 (A) (B) (C) (D) (E)	94 (A) (B) (C) (D) (E)
15 (A) (B) (C) (D) (E)	35 (A) (B) (C) (D) (E)	55 (A) (B) (C) (D) (E)	75 (A) (B) (C) (D) (E)	95 (A) (B) (C) (D) (E)
16 (A) (B) (C) (D) (E)	36 (A) (B) (C) (D) (E)	56 (A) (B) (C) (D) (E)	76 (A) (B) (C) (D) (E)	96 (A) (B) (C) (D) (E)
17 (A) (B) (C) (D) (E)	37 (A) (B) (C) (D) (E)	57 (A) (B) (C) (D) (E)	77 (A) (B) (C) (D) (E)	97 (A) (B) (C) (D) (E)
18 (A) (B) (C) (D) (E)	38 (A) (B) (C) (D) (E)	58 (A) (B) (C) (D) (E)	78 (A) (B) (C) (D) (E)	98 (A) (B) (C) (D) (E)
19 (A) (B) (C) (D) (E)	39 (A) (B) (C) (D) (E)	59 (A) (B) (C) (D) (E)	79 (A) (B) (C) (D) (E)	99 (A) (B) (C) (D) (E)
20 (A) (B) (C) (D) (E)	40 (A) (B) (C) (D) (E)	60 (A) (B) (C) (D) (E)	80 (A) (B) (C) (D) (E)	100 (A) (B) (C) (D) (E)

18606243R00073

Made in the USA
Middletown, DE
12 March 2015